DISCARDED
From Nashville Public Library

D0810206

American Alligator

UNIVERSITY PRESS OF FLORIDA

Florida A&M University, Tallahassee
Florida Atlantic University, Boca Raton
Florida Gulf Coast University, Ft. Myers
Florida International University, Miami
Florida State University, Tallahassee
New College of Florida, Sarasota
University of Central Florida, Orlando
University of Florida, Gainesville
University of North Florida, Jacksonville
University of South Florida, Tampa
University of West Florida, Pensacola

American Alligator

Ancient Predator in the Modern World

Kelby Ouchley

University Press of Florida

Gainesville

Tallahassee

Tampa

Boca Raton

Pensacola

Orlando

Miami

Jacksonville

Ft. Myers

Sarasota

Frontispiece: Photo by Burg Ransom.
All color photos are by Burg Ransom.

Copyright 2013 by Kelby Ouchley
All rights reserved
Printed in the United States of America on acid-free paper

This book may be available in an electronic edition.

18 17 16 15 14 13   6 5 4 3 2 1

Library of Congress Cataloging-in-Publication Data
Ouchley, Kelby, 1951–
American alligator : ancient predator in the modern world / Kelby Ouchley ; color photographs by Burg Ransom.
p. cm.
Includes bibliographical references and index.
ISBN 978-0-8130-4913-7 (alk. paper)
1. American alligator. 2. Alligators—North America. I. Ransom, Burg. II. Title.
QL666.C925O93 2013
597.98′4—dc23   2013020092

University Press of Florida
15 Northwest 15th Street
Gainesville, FL 32611-2079
http://www.upf.com

For field biologists then and now:

The old ones who revealed the pieces;

And the new ones who assemble the puzzles

Contents

Introduction

I suppose it could be said that because of an alligator my father almost lost his life when I was nine years old. While inspecting a pipeline from a helicopter in the vast Louisiana coastal wetlands, he noticed a female alligator near a nest and asked the pilot to swoop in for a closer look. For reasons that were never made clear to me, the helicopter crashed at the nest site. Miraculously, both my father and the pilot survived. My memories of the incident are linked to old color photographs of the plastic-bubbled whirlybird upside down and mostly submerged in a maidencane marsh.

This event only fueled my budding curiosity of all things associated with the natural world. I am sure my persistent questions led to the wet burlap sack that was dropped at my feet soon after the helicopter accident. The bag croaked. To my delight it contained a yearling alligator, which other than a collie pup for a few months, became my first real pet. Soon afterward, another sack arrived with four brightly banded hatchlings, and a part of my adolescent life became devoted to alligator husbandry for several years. Over time, and in the manner of osmosis, I unwittingly absorbed a good deal of basic biology concerning the reptiles in my tanks. They loved crawfish and dragonflies, but nothing could entice them to eat during winter months. Their senses were such that they alerted to my presence even at a distance. They emitted a pungent musk when riled, and the larger one did not hesitate to attack the small ones given the opportunity. Strange anatomical features were revealed in the form of nictitating membranes and palatal valves. I was only the latest victim of a species that had been beguiling humans for thousands of years.

My family moved inland from the coast to a hill on the edge of a bottom-land hardwood swamp dissected by a serpentine bayou. Although the habitat was suitable for alligators, they had long since been extirpated from the immediate area—until, that is, one summer day when "Old Grandpap" swam up from the Ouachita River to our swimming hole under the Bayou D'Arbonne bridge. We perpetuated the myth that he must be a hundred years old. He was almost as long as my 12 foot bateau and he was a magnificent creature. One of his eyes was milky white and sightless, likely a result of a gunshot. I could approach him on his blind side, paddling quietly to within a few feet, as he floated midstream in deep water. When he detected me, the explosive power of his diving escape was beyond anything I had ever experienced in the animal world. More than a half century later I think that continues to be the case.

I released my pet alligators in the swamp, Old Grandpap moved on, and so did I. After graduate school on the western edge of alligator range, I began a 30-year career as a wildlife biologist and manager of National Wildlife Refuges (NWRs) with the U.S. Fish and Wildlife Service. Soon I was getting paid to work with alligators from time to time. At Holla Bend NWR on the Arkansas River, another area that had lost the species, I helped reintroduce alligators. On this detail I learned that alligators do not travel well. While transporting a load up from the Louisiana marshes, I noticed in the rearview mirror a six-footer, untied and untaped, slither over the tailgate of my truck. As a result I found myself sitting astraddle of the offender on the yellow stripe of a road somewhere in the piney hills near Hamburg, Arkansas. I could barely hold him in place; I certainly could not get him back into the vehicle alone, and the passing pulpwood truck driver would have none of it. During this same large relocation effort, involving hundreds of alligators and several states, a friend watched an 11-footer escape his bonds Houdini-like, rise up over the back of his truck cab, devour his radio antenna, and depart the vehicle heading back south down I-49 near Jackson, Mississippi.

I moved back to the marshes and prime alligator habitat at Lacassine NWR in southwest Louisiana. There my most intensive work with alligators occurred as we began a capture/tagging project to determine the parameters of a population that had not been hunted in at least 50 years. Many exciting nights were spent in an airboat catching and collecting data on alligators of all sizes. The duties were not all glitch-free, as noted in my 1981 field diary entry: "On night of 5 June, I motored to northeast spillway of Lacassine Pool in Whaler. Carried pirogue to photograph in Pool. Photographed alligators, raccoon, pig frogs, lotus blooms, green snake, and marsh. Ran aground in

Whaler on return trip and had to paddle and pole pirogue 5 miles home. Arrived 2 a.m."

From Lacassine I was transferred to the vast Tensas Swamp, where it is very likely that the last ivory-billed woodpecker in the world, a female, flew into the abyss of extinction. When I arrived, alligators from her realm still nested along the cypress sloughs and brakes where she once foraged. Another field diary entry from September 1983 reads: "To Lost Brake to check on gator poaching rumor in a.m.; found gator nest on dam; brought 7 eggs home but most hatched before I got there." In that case the female alligator was thought to have been killed.

My job description required that I have collateral duties as a federal game warden, and law enforcement incidents pertaining to alligators surfaced from time to time in my area of responsibility. They ran the gamut of creative lawbreaking from the poacher who decided to pressure wash his illegally killed alligator at a self-serve carwash at 3:00 a.m. to the entrepreneurs who were allegedly trading baby alligators to a pet store for cocaine.

After other duty stations, all of them encompassing alligator habitat, I am living back on the edge of the swamp that Old Grandpap visited. In a pleasant and improbable turn of events, alligators are common here now. When I was given the hatchlings in 1960, I did not understand that the survival of their kind as a viable component of wetland ecosystems throughout their historical range was doubtful. Perhaps it was their scarcity then that enticed my father to take a closer look at the nest from the helicopter. Or maybe it was something else. As an apex predator, a carnivorous species that lurks at the top of the food web, the alligator toys with our psyche not unlike the mythological Sirens. Once exposed, we are vulnerable. For this reason I have made this effort to tell their story, an up-to-date account of the long winding trail of the American alligator.

Part I

Natural History

Alligator, alligator, lives in the swamp;
Alligator, alligator, chomp, chomp, chomp.

1

In the Beginning

Evolution/Paleontology

The trailhead for modern crocodilians can be found in the Late Cretaceous period about 80 million years ago. If it were possible to go back in time and orbit planet Earth in a spaceship, the geography of our world at that moment would be foreign to us. Then as now, the continents were adrift, but having only recently broken apart, North America was much closer to Europe. Africa was nearer to South America and farther from Eurasia. India was attached to Madagascar, while Australia abutted Antarctica. The Atlantic Ocean was much narrower, the Pacific much wider. A sea that we now call the Tethys Ocean comprised the Indian and Antarctic Oceans. The entire continent of North America was divided by a large inland sea (Western Interior Seaway) that ran the length of the Rocky Mountains' Front Range. Sea levels were higher in the Late Cretaceous than at any other time in Earth's history—about 650 to 800 feet higher than at present. All of Florida and the gulf coast were submerged. Indeed, almost all of the modern range of American alligators was beneath a shallow sea.

Geologists believe that seafloors were spreading faster then, which resulted in a period of active volcanism around the world. These thermal commotions may have in turn affected the climate that was warmer and more humid than today. Plant fossils reveal that tropical to subtropical conditions existed in regions far from the equator, and a temperate climate extended to the poles. Amazingly, dinosaurs thrived in Antarctica.

It was a time when dinosaurs of such diversity as to bend the imagination ruled the land, giant marine reptiles like long-necked plesiosaurs and fish-shaped ichthyosaurs reigned in the seas, and flying reptiles with wingspans that measured three times the length of today's longest alligator patrolled the skies. However, not all life then would seem alien to us. Mammals, fish, snakes, and lizards, all with modern characteristics, appeared first in the Cretaceous period. Even plants had evolved a contemporary look as angiosperms (flowering plants) out-competed more primitive forms. We would recognize magnolias, sycamores, willows, figs, and some herbaceous plants of the time.

Such an alien world could mislead one to think that all was unordered chaos. It was not. The same basic ecological mechanisms that drive ecosystems in the Amazon basin and Okefenokee Swamp today were at work 80 million years ago. In order to meet needs of food, water, shelter, and space, organisms had adapted to specific habitats. Each had a niche or "occupation" that defined its role in the ecosystem. All was fueled by the energy of our nearest star, which flowed first into green plants and then upward through the food webs to peak in top predators. Even then, death and decomposition was only a feedback loop in the cycle of life.

A basic tenant of evolution is that life forms change over time. Those species that develop traits, either by gene mutation or learning, that make them more likely to survive and pass those characteristics on to their offspring tend to march greater distances down life's unpredictable paths. Others less "fit" for their environment wander into the abyss of extinction.

Enter then onto this strange world stage of the Late Cretaceous a creature that might be the last common ancestor of modern alligators, caimans, crocodiles, and the gharial. Not an individual animal but rather a functional species whose offspring changed over thousands of generations into the modern lineage of crocodilians, he has been called "shieldcroc" for the shield-like bony plate on his head. Adults exceeded 33 feet in length including a 6½-foot-long head with small teeth and weak jaws. Shieldcroc's fossils were first discovered in freshwater deposits of present-day Morocco, a region some paleontologists believe to be the evolutionary home of modern crocodilians.

Shieldcroc, of course, did not appear on the roadside unannounced. The first clearly recognizable crocodilian-like fossils date back to the Triassic period, over 200 million years ago, about the time dinosaurs were showing up. Along with the dinosaurs, pterosaurs (flying reptiles), and birds, ancestral crocodilians were a branch of the archosaurs that had common anatomical characteristics including teeth that were set in sockets. With a firm anchor-

"Shieldcroc," a Late Cretaceous ancestor of modern alligators. Painting by Henry P. Tsai.

ing, teeth are less likely to be torn loose during feeding. (It should be noted that the bird line of archosaurs eventually became toothless.) Archosaurs also shared unique skull openings in front of the eyes and in the jaw that reduced the weight of large heads. A third trait common to the group was the presence of a special knob on their thigh bones for muscle attachment. This protuberance allowed the development of large muscles that enabled the creatures to stand and walk with legs tucked under the body instead of slithering about on splayed limbs. Faster locomotion was a result. Most of those in the early proto-croc branch were the size of dogs. Scientists believe that they were likely terrestrial (land-dwellers) for a good 20 million years before they invaded the swamps, rivers, and seas. Some were even vegetarians. Examples include the one-foot long *Erpetosuchus* found in Scotland and *Doswellia* from eastern North America.

Because archosaurs were the common ancestors of birds and crocodilians, birds are now the closest living relatives of alligators and their kin. When the completed mitochondrial genome of the alligator was sequenced in the 1990s, it was determined that alligators evolved faster than other cold-blooded vertebrates and birds. This challenged long-held beliefs that warm-blooded animals (including birds) evolved at a faster rate. The genome project also indicated that the avian and crocodilian branches of archosaurs split about 254 million years ago.

About 150 million years ago at the beginning of the Jurassic period, our subjects moved toward an aquatic existence. We would recognize them as ancestors of our modern alligators. Their bodies grew long and streamlined, their skulls became flattened and armed with fearsome dentition and crushing jaws. Their legs splayed once more. Continuing to evolve for another 50 million years in direct competition with dinosaurs, some species joined the race for dominance by growing to enormous sizes. Along the American coastlines and within the Western Interior Seaway, *Deinosuchus*, approaching 40 feet in length and weighing 8½ tons, was a beast to be reckoned with even by the most formidable of dinosaurs.

Thus we now arrive back (or forward as the case may be) in the Late Cretaceous to the time of shieldcroc. Some researchers believe that salt intolerance may have dispersed its evolving offspring along land bridges. For the first time the fossil record begins to reveal a clear distinction between crocodiles and alligators. Diversity was manifest. *Brachychampsa montana*, an early alligator, had morphological adaptations including specialized teeth that indicate it ate mostly turtles. Alligators and caimans also separated from each other in North America then and spread to South America before the Paleocene epoch (about 65 million years ago) when these two continents were disconnected.

An organism's ability to progress along life's road depends on its success at overcoming obstacles at different levels. At the most basic, it must find adequate food, shelter, and mates to carry on during a lifetime. On a much grander scale it must surmount infrequent cataclysmic barriers that threaten the species as a whole. These last are often termed "extinction events," periods when the abundance and diversity of life on Earth plummeted. A glance in the rearview mirror of modern alligators' pedigree exposes three such bottlenecks of catastrophe. The Permian-Triassic event (250 million years ago) was the largest when 96 percent of all marine species and 70 percent of all land species died out. All of the archosaurs lines managed to survive, perhaps because of well-developed hip muscles that allowed efficient mobility. In the Triassic-Jurassic event (200 million years ago), many archosaurs became extinct, but pterosaur, crocodilian, and dinosaur lines made it through. The most recent Cretaceous-Tertiary event (65½ million years ago), when 75 percent of all species became extinct, eliminated pterosaurs and most non-avian dinosaurs. Once again crocodilians and their bird cousins managed to avoid the wreck.

Causes of extinction events vary and are often poorly understood. Explanations include asteroid impacts, erratic global warming and cooling, and unusual volcanic eruptions. Diversity among crocodilians reached a peak eons ago. Only 23 species remain. That the alligator lineage survives today in a drastically altered world speaks to their degree of fitness and ability to adapt to environmental changes over 200 million years—not bad for a critter whose brain weighs less than half an ounce.

Names and Places

Taxonomy

The accepted common name for the subject of this book is American alligator. Alligator is an Anglicized term from the Spanish "el lagarto" (lizard), whose origins can be traced to alligator encounters by early Spanish explorers in Florida. In a twist, the transplanted French Acadians took another Spanish word that refers to crocodilians, "cocodrilo," and transformed it to "cocodrie"—an alligator expression that survives in geographic place names of south Louisiana (for example, Bayou Cocodrie).

Humans have been putting common names on plants and animals ever since the first cave man found it advantageous to convey to his mate the difference between a cave cricket and a cave bear. As our species developed culturally and interactions between groups who spoke different languages increased, the matter surfaced again. Hypothetically, the folks on the other side of the mountain were peddling rugs from an animal they called moose, but which were known to the prospective buyers as elk. It was a problem of conflicting common names, and it got worse when scientists came along. A Swedish biologist named Carolus Linnaeus began working on the issue in the 1740s and developed a system to give a two-part Latin name to every plant and animal on the planet. Using Linnaeus's system, the renowned French herpetologist, François Marie Daudin, officially described and named the American alligator in 1802. Now, when scientists in Japan, Somalia, and Brazil read this book about *Alligator mississippiensis* by an author from Rocky Branch, Louisiana, they know I am referring specifically to the American alligator. We call the Latin name of a species the "scientific name."

Taxonomy is the science of classifying organisms. Nomenclature associated with taxonomy was often a bit messy as early naturalists tried to sort out who was kin to whom. Names changed as new insights were revealed in the field of biology. (Today the shuffling occurs primarily as a result of the new discipline of DNA analyses that can determine genetic commonalities of various organisms.) Precedent went to names that were published first, but such information was not as readily available to researchers in different parts of the world as is the case now. It was not uncommon for a name to change when someone discovered an earlier published description of a plant or animal. This happened with the American alligator when another Frenchman, Georges Cuvier, described the species in 1807 and called it *Crocodilus lucius*. It was later determined that this was the same reptile previously described by Daudin in 1802 and named *Crocodilus mississipiensis*. As a result, Daudin gets credit for naming the species. Ironically, Cuvier coined the genus *Alligator* into which the species was later tossed when it was decided that the creature is suitably distinct from crocodiles. The species *mississipiensis* was emended to *mississippiensis* in 1842 on the basis that the name refers to the Mississippi River.

Linnaeus's taxonomic system, which is still in use today although slightly modified, went beyond just naming all life forms by creating seven major divisions (taxa) that tend to show degree of kinship between organisms. The eighth category (Domain) was recently added. Groupings are:

Domain → Kingdom → Phylum → Class → Order → Family → Genus → Species

The classification becomes more specific as one progresses from domain toward species. All animals share the same domain; fewer share the same family. Likewise, two creatures in the same genus are much closer related than

Table 1. Comparison of the Classification of the American Alligator and American Crocodile

	American Alligator	American Crocodile
Domain	Eukarya	Eukarya
Kingdom	Animalia	Animalia
Phylum	Chordata	Chordata
Class	Reptilia	Reptilia
Order	Crocodylia	Crocodylia
Family	Alligatoridae	Crocodylidae
Genus	*Alligator*	*Crocodylus*
Species	*mississippiensis*	*acutus*

two who only share the same class. Individuals of the same species are genetically distinct and capable of interbreeding. Above the species level, organisms generally do not interbreed in the wild. An example would be gray and fox squirrels that have the same genus, *Sciurus*, but different species names (*carolinensis* and *niger*, respectively). Even though they often share the same habitat, interbreeding is rare. It should be noted that many subdivisions (such as subphylum and subspecies) exist within the eight principal levels.

Table 2. Taxonomic Chart of Living Crocodilians

Family Alligatoridae
    Genus *Alligator*
        American Alligator *(Alligator mississippiensis)*
        Chinese Alligator (*Alligator sinensis*)
    Genus *Paleosuchus*
        Cuvier's Dwarf Caiman *(Paleosuchus palpebrosus)*
        Smooth-fronted Caiman (*Paleosuchus trigonatus*)
    Genus *Caiman*
        Yacare Caiman (*Caiman yacare*)
        Spectacled Caiman (*Caiman crocodilus*)
        Broad-snouted Caiman (*Caiman latirostris*)
    Genus *Melanosuchus*
        Black Caiman (*Melanosuchus niger*)
Family Crocodylidae
    Genus *Crocodylus*
        American crocodile (*Crocodylus acutus*)
        Slender-snouted crocodile (*Crocodylus cataphractus*)
        Orinoco crocodile (*Crocodylus intermedius*)
        Freshwater crocodile (*Crocodylus johnsoni*)
        Philippine crocodile (*Crocodylus mindorensis*)
        Morelet's crocodile (*Crocodylus moreletii*)
        Nile crocodile (*Crocodylus niloticus*)
        New Guinea crocodile (*Crocodylus novaeguineae*)
        Mugger crocodile (*Crocodylus palustris*)
        Saltwater crocodile (*Crocodylus porosus*)
        Cuban crocodile (*Crocodylus rhombifer*)
        Siamese crocodile (*Crocodylus siamensis*)
    Genus *Osteolaemus*
        Dwarf crocodile (*Osteolaemus tetraspis*)
    Genus *Tomistoma*
        False gharial (*Tomistoma schlegelii*)
Family Gavialidae
    Genus *Gavialis*
        Gharial (*Gavialis gangeticus*)

In addition to American alligators, 22 species of crocodilians exist today, and all are found within the order Crocodylia. Separation begins at the family level. American alligators share family Alligatoridae with the Chinese alligator and six species of caimans. Fourteen species of crocodiles are in family Crocodylidae, and the gharial is the sole surviving species in family Gavialidae. Family groups can be easily distinguished by head shape among other characteristics.

Range, Habitat, and Abundance

Factors that determine where American alligators can live and sustain healthy populations include food, water, and temperature on a year-round basis. The necessary parameters of each can be found in a variety of habitat types, which together make up the alligator's range. As a result of adaption to survive in areas with temperate climate, American alligators are the most northerly distributed of all crocodilians in the world. Breeding populations exist as far north as 35° latitude in the marshes and rivers of coastal North Carolina. Within that region of the United States having a survivable climate for alligators, the quality and quantity of wetlands determined their historical presence or absence. After disappearing from many areas where they were once found, usually because of human exploitation, alligators in recent years have returned to occupy most of their known historical range.

The present alligator range is from coastal North Carolina south to Florida and westward to central Texas and southeastern Oklahoma. Their status in each state with historical populations follows:

North Carolina. Alligators are found in most coastal counties of North Carolina. From Alligator River NWR in Hyde County south to Brunswick County, they live in tidal estuaries, marshes, swamps, streams, canals, ponds, and lakes. The largest populations live in the southernmost counties, but others also thrive in the lakes of Croatan National Forest in Jones, Craven, and Carteret counties. The North Carolina Wildlife Resources Commission has no estimate of the size of the alligator population.

South Carolina. Alligators are found below the fall line (the boundary that runs between the upland sandhills and Piedmont geographic regions) and throughout the Coastal Plain of South Carolina. Most are in the coastal wetlands with highest nest densities located in the ACE Basin, an estuary of biological importance drained by three major rivers. Some of the largest populations are in long-abandoned rice fields now managed as waterfowl hunting areas. Suitable habitat declines as one moves inland from the coast. The estimated alligator population in South Carolina is a minimum of 100,000.

Georgia. In Georgia alligators occur in the Coastal Plain south of the fall line that roughly connects Augusta, Macon, and Columbus. They are common in coastal marshes, swamps, rivers, streams, and lakes throughout the region and are abundant in the Savannah River and its associated wetlands. Okefenokee NWR in south Georgia is estimated to harbor 10,000–13,000 alligators. The estimated state population is 200,000.

Florida. Alligators are found in every Florida county from the Florida Keys northward to the Okefenokee Swamp on the Georgia border. The state's wetlands are diverse and include brackish and freshwater marshes, large natural lakes, inland swamps, cypress ponds, and a plethora of rivers and streams. The vast Everglades of south Florida was once prime alligator habitat but has been seriously degraded by drainage projects and agriculture in the last hundred years. Restoration is ongoing. Some of the best remaining habitat occurs in the large lakes of central Florida. Almost three million acres of alligator habitat are in public ownership and subject to some degree of protection. The alligator population in Florida is estimated at 1.3 million.

Alabama. Historically, it appears that reproducing populations of alligators in Alabama were restricted to the Gulf Coastal Plain in the south half of the state. Field biologists report that their present range has expanded northward in the Piedmont and Valley and Ridge physiographic regions to now encompass the southern two-thirds of the state. Five rivers flow through the Gulf Coastal Plain with the Alabama River being the largest. Wetlands associated with these rivers and especially the coastal and inland waters near the Mobile Delta harbor the most alligators. The population is stable to increasing in occupied habitat and estimated at 25,000–35,000.

Mississippi. Alligators are found throughout Mississippi except for the Tombigbee Hills region and several northern counties where winter temperatures likely limit their habitation. Biologists estimate that there are about 408,000 acres of alligator habitat in the state with coastal Jackson County having the most at 57,000 acres. Most alligators occur in Jackson County and nearby Hancock County. Rankin County in the central part of the state encompasses Ross Barnett Reservoir, another area with abundant alligators. Mississippi's estimated alligator population is 32,000–38,000.

Louisiana. As in the case of Florida, alligators are found throughout Louisiana in a variety of habitats including marshlands, swamps, bayous, rivers, lakes, ponds, and canals. Tragically, Louisiana contains about 40 percent of the nation's wetlands and experiences 90 percent of the coastal wetland loss in the lower 48 states. Losses are a result of human activities including levee building and oil and gas development that have impaired the natural deposition of sediments from the Mississippi River and have enabled saltwater intrusion, and conversion of wetlands to agriculture. These activities have exacerbated the impacts of natural processes such as hurricanes. The current rate of loss is 25–35 square miles per year. An estimated 4½ million acres of alligator habitat remains in Louisiana with more than 3 million acres located in coastal freshwater and brackish marshes. Cypress-tupelo swamps comprise 750,600 acres of habitat, 207,000 acres are the Atchafalaya Basin swamp, and numerous lakes contribute 32,105 acres. The alligator population in Louisiana is estimated at almost 2 million.

Texas. Alligators are found in about 120 counties of east Texas beginning at the Sabine River and west to near I-35. Highest concentrations occur in the Gulf Coastal Plain and especially the coastal marshes along the Gulf of Mexico south to the Rio Grande River. In general, the quality of inland habitat declines in a westerly direction. Within the historical range, several large, man-made reservoirs provide suitable habitat that did not exist naturally. Also, with the construction of new lakes and water diversion projects, alligators now occur in some areas outside their historical range. The Texas alligator population is officially estimated at about 500,000 with more than half found in the upper coastal counties of Jefferson, Chambers, and Orange.

Arkansas. As recently as 1973, alligators in Arkansas were restricted to three counties in the southwestern corner of the state—Hempstead, Miller, and Lafayette. At that time the population was estimated at 1,900. A year earlier the Arkansas Game and Fish Commission began a 12-year project to restock alligators in their assumed historical range, which is thought to include the Coastal Plain and the Mississippi Delta physiographic region south of a line between Little Rock and Memphis. Today, alligators are found beyond this area as far northwest as Holla Bend NWR on the Arkansas River, and as far northeast as the St. Francis River near Paragould. The highest densities are in the southwestern corner near Millwood Reservoir and the southeastern corner around Arkansas Post National Monument. The state population is estimated at 3,000–4,000.

Oklahoma. The status of alligators in Oklahoma is unclear as it pertains to their historical range and the origin of the existing population. Some historical records indicate that alligators were native to the state in small numbers, but were likely extirpated prior to 1900. The first nest was not documented in the state until 2005 at Red Slough Wildlife Management Area in McCurtain County. Most alligators today are found in the extreme southeastern counties of McCurtain, Choctaw, and Bryan. The current population is believed to be the result of released pet alligators and immigration from Louisiana and Arkansas. The size of Oklahoma's alligator population is unknown.

. . .

Many records exist of alligators found far outside of their natural range, with documented sightings in California, Colorado, Indiana, Missouri, New York, New Jersey, Pennsylvania, Tennessee, Virginia, and West Virginia. In all cases their presence can be explained as intentional or unintentional releases of captive animals, or aberrant immigration. Usually they survive no longer than a few months, but in rare instances individuals may live several years if winter temperatures are mild. In Alabama, Arkansas, and perhaps Texas, breeding populations of alligators may exist in areas up to 100 miles outside of their historical range. No breeding populations are known to exist at greater distances, despite such urban myths as the population of alligators said to inhabit New York City's sewer system.

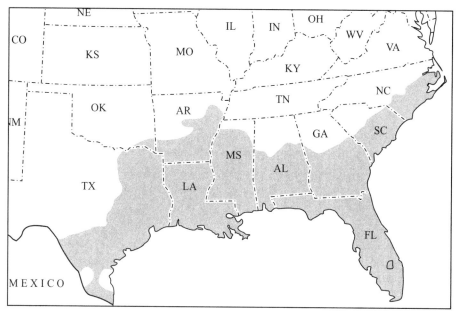

Range of the American alligator.

3

"Their Toes Are Five in Number"

Their toes are five in number on the anterior feet, and four on the posterior; their sharp and conical teeth are arranged in a single series in each jaw; their tongue is flat, fleshy, and closely attached almost to its very edge; and their bodies are clothed with large, thick, square scales, the upper of which are surmounted by a strong keel, those of the tail forming superiorly a dentated crest, double at its origin.

—Description of an alligator kept in the Tower of London menagerie in 1829

Description

Alligators are reptiles in the taxonomic class called Reptilia. Members of this group have common characteristics. All are ectothermic (cold-blooded) and have backbones. Most have four limbs (except snakes, which have four-limbed ancestors), reproduce by laying eggs with shells (except, again, for some snakes), and have bodies covered in scales or scutes. Within class Reptilia, alligators are placed in the subdivision known as order Crocodylia and are referred to as crocodilians. Members of this group have similar anatomical traits and include alligators, crocodiles, caimans, and the gharial. Alligators differ from crocodiles by having a broader snout and an upper jaw that overlaps teeth in the lower jaw. The gharial has a long, slender snout. Alligators most closely resemble caimans that live in Central and South America. American alligators grow larger than their closest relative, the Chinese alligator, which rarely exceeds seven feet in length.

The lizard-like body shape of alligators is well known to most people. The legs are stubby, powerful, and capable of driving the animal to a gallop for short distances. The front feet have five clawed toes and the back feet have four. A long, muscular tail flattened on the vertical axis propels the alligator during swimming. The snout is generally broad, but the exact proportion seems to be a genetic trait that varies widely. Captive populations tend to have broader snouts, an indication that diet also impacts snout shape. Nostrils and eyes protrude above the plane of the snout so that the alligator can submerge with only these organs above the level of the water. (Many other aquatic animals have similar eye/nostril arrangements, the hippopotamus being a notable example among mammals.) Eye color ranges from green to silverish.

Tough, leathery skin adheres to bony plates called osteoderms on the back and tail. Osteoderms form during embryonic development as connective tissue transforms into bone-like structures. All crocodilians are armored with osteoderms, and caimans even have them on their undersides. In addition to serving as a defensive shield, these highly vascularized components of the exoskeleton function as heat-exchangers to enable critical warming and cooling as necessary. Skin on the underside of alligators is composed of embedded scales somewhat quadrangular and thinner than those on the back. Scale patterns vary between populations in the eastern and western parts of the species' range in different ways including the total number of scales and number of rows of scales.

The upper surface of adults is uniformly dark while the underside is creamy white. White speckling occurs on the jaws of some populations. Juveniles are camouflaged with yellow cross-bands on their black back and tail. In hatchlings the cross-bands are white, and research with lab animals shows that the number of bands is affected by incubation temperature. Alligators that incubate at warmer temperatures have more stripes.

Two rare genetic manifestations of alligator skin have attracted thousands of visitors to some zoos and other reptile exhibits in recent decades. The sensations began when 18 white hatchlings were found on a nest near Houma, Louisiana, in 1987. They were not albinos but rather leucistic alligators. An animal with this condition has defective skin cells that are incapable of making any color pigments on all or part of its body (except for the eyes). In contrast, albinism is a genetic trait that inhibits the production of only the dark pigment melanin. Another difference between albino and leucistic animals is eye color. The eyes of albinos appear red because the internal blood vessels are not masked by a dark pigment, but those of leucistics are normal. Some of the captive Houma alligators are still alive and exceed 10 feet in length.

Albino alligators have surfaced only slightly more often than leucistic ones. The acclaimed herpetologist Raymond Ditmars wrote in 1908, "There is an albinistic specimen [of alligator] living in the New York Zoological Park." An alligator farmer collecting wild eggs near Myrtle Grove, Louisiana, in 1991

Albino alligator. Photo by Mila Zinkova.

found a clutch that hatched seven albinos. Over the next several years he acquired albinos from two nests in the same area each season. His conclusion was that there were at least two local females and one male alligator that carried the recessive gene for albinism. As recently as 2012, an albino alligator was exhibited at the National Aquarium in Washington, DC. Curators there stated that, "Fewer than 100 of these extraordinary species exist worldwide due to the many environmental challenges that they face." White alligators, whether albinos or leucistics, have no chance of extended survival in the wild. Their skin is sun-sensitive, and the stark coloration highlights them as easy targets for abundant predators.

In the imaginations of most folks, teeth define alligators—with good reason—their dentition is formidable in appearance and quantity. Lions and tigers have 30 teeth including 4 fearsome canines; alligators have 74 to 80 and all are designed not to chew, but rather to stab, pierce, and grip prey. Alligator teeth are pointed, hollow, and set in bony sockets (referred to as thecodont dentition). Each tooth has a new one growing inside ready for immediate use when the old one is lost. Tooth replacement in all crocodilians is continuous throughout life or at least until they are very old. Young rapidly growing al-

X-ray CT scan of alligator skull. Courtesy of DigiMorph.org.

ligators replace teeth much faster than adults. Some obviously old individuals are almost toothless. Alligators conceivably wear through 2,000 to 3,000 teeth in a lifetime.

Deadly teeth are of little value without powerful jaw muscles to support their work. A prominent New Orleans physician studying the dentition of large live alligators in 1846 after "the mouth was opened, and was retained so by strong levers, in order to facilitate the experiments, and to prevent the crushing of the arm," determined that "the crushing power of the jaws is verticle, not lateral or grinding." The force of an alligator's bite has been measured and as might be expected is linked to the size of the animal. One very large (over 13 feet), wild Florida alligator bit down on a measuring instrument with a force of 2,960 pounds. For comparison, others have calibrated the bite force of dusky sharks at 330 pounds, lions at 940 pounds, hyenas at 1,000 pounds, and humans at 170 pounds. Power is only in the downward closure of the jaws as a man can easily hold a large alligator's mouth shut. The hollow, muffled sound of a huge crocodilian smashing its jaws closed is singular in the natural world. I hear it as a giant, empty suitcase slamming shut.

As a boy feeding baby alligators, I remember being intrigued by the fact that I could not see down the throats of my gaping pets. The reason was because the back of the tongue is shaped into a structure called the palatal valve. When closed, this mechanism allows an alligator to open its mouth underwater without choking. The internal opening of the nostrils enters the throat cavity behind the valve to permit breathing with just the tip of the nose exposed. Each of the two external nostrils on the end of the snout is equipped with an additional protective valve in the form of a flap that closes when the alligator submerges, or in the case of the curious boy, when quickly touched with the tip of a finger.

Crocodiles and the gharial have glands on their tongue that excrete excess salt from the body. Alligators and caimans have nonfunctional remnants of the same glands. This reinforces the idea that the ancestors of modern crocodilians lived in marine environments.

Swallowed food passes down the throat through the esophagus to the stomach. Small stones or other hard objects (gastroliths) are often found in the stomach, and it has been suggested that, in the fashion of a gizzard, they aid in digestion. Some think these objects may also serve as ballast. But neither idea has been proven. Creative ideas regarding alligator gastroliths have been around since at least 1806 when it was noted in a Philadelphia medical journal that, "The alligator, having previously swallowed a number of pine-knots, retires to his hole, where he remains in a torpid state during the severity of the winter. If killed at this season, these knots are found highly polished by their trituration one against the other in the animal's stomach. . . . I apprehend that these substances, when taken in by the animal, act in some measure by keeping up a certain degree of action in its stomach, and consequently in every part of the system, and thereby prevents the death of the animal, which might otherwise be destroyed by the long continued application of cold."

Digestive juices in the smaller part of the stomach have been reported to be the most acidic of any animal. In unique fashion, oxygen-depleted blood with high levels of carbon dioxide can be routed to the stomach to produce the acid necessary to digest bones, shells, scales, and other hard items. Some researchers working with stomach contents have experienced a severe burning sensation when digestive juices came in contact with their unprotected skin. From the stomach, food remnants progress to the intestines comprised of a long, coiled small intestine, a wider rectum, and the cloaca. The resulting feces are voided through the vent or cloacal opening.

Crocodilians are the only reptiles to have four-chambered hearts. Others have three. Like birds and mammals, alligators have two ventricles and two atria. Unlike birds and mammals, some mixing of the blood occurs between chambers through an opening in the atria called the foramen of Panizza. The blood chemistry of alligators is somewhat similar to mammals and birds. Research indicates that components of alligator blood have powerful antibiotic qualities.

Recent study has questioned the long-held assumption that crocodilians exhibit two-way airflow through the lungs like mammals. It was believed that a diaphragm-like organ and other muscles worked to increase or decrease lung size that in effect resulted in the animal inhaling or exhaling air. New tests indicate that air-flow in alligator lungs is one-way, just as in birds. Instead of using gas-exchanging alveoli found in mammals, alligators breathe with the aid of tubes called parabronchi where air flows through the lungs for eventual escape through the windpipe. If accurate, this complicated physiological procedure has evolutionary significance in that it may have helped crocodilian ancestors survive at a time when oxygen levels on Earth were much lower than today (12 percent versus 21 percent).

The old adage that an alligator has a brain the size of a walnut is, in the case of adults, fairly accurate. Most reptiles have relatively small brains, but within this group it is more intricate in crocodilians. The alligator's brain lies in the top center of its skull protected all around by thick bone. In this position it can also warm quickly for maximum efficiency when the alligator basks. Alligators have a true cerebral cortex and well-developed optic and olfactory regions that validate the importance of sight and smell. One of the most amazing statistics of nature must be that the whole of the complexity of alligator being is driven by an organ that weighs less than a third of an ounce (8 to 10 grams).

We humans describe the world that we see around us as our environment. Our perception of this world is unique in that none of the other living creatures on the planet share our experience. Each species has its own awareness of what we call the environment. The term for what an animal perceives is umwelt, or self-world. There can be as many different umwelten in a particular environment as there are kinds of animals. The umwelt for any creature is dependent upon the types of sensory receptors that it possesses and its capability to process stimuli of those receptors. By considering the alligator's sensory organs we can only imagine its worldview.

The source of one remarkable alligator sense is found in the tiny, dark bumps sprinkled across the sides of its jaws and mouth. When alligators hunt at night these organs are situated just at the air/water interface. Recently named "dome pressure receptors," the bumps are extremely sensitive to disturbances of the water surface. When researchers blocked other sensory organs (eyes, ears, and nose) in a controlled experiment and let a single droplet of water fall into the pool, the alligators instantly lunged and snapped in the direction of the disturbance. If the receptors were afterward blocked with a masking coating, they did not respond as before. The bumps

"Dome pressure receptors" on the side of an alligator's jaw. Photo by Burg Ransom.

are filled with nerve bundles, although the exact workings of the process are unknown. Instead of responding to movement of the ripples, alligators may be sensing pressure changes in the water's surface—a novel tactic to detect potential prey.

Animals that spend time above and beneath the surface of water face special challenges when it comes to the sense of hearing. When above water, alligators process sound as it enters the outer ear and passes through to stimulate nerve hairs in the inner ear. Below the surface, the inner ear likely receives sound stimuli through bones of the skull. Alligators, when tested, were able to hear across a greater frequency range in air than underwater. Indeed, they were able to hear as well as many land-dwelling species including birds. Peak sensitivity seems to be in the lower frequencies in the same range as adult vocalizations. Underwater, they were still able to hear quite well although the range of detected sound was smaller. Not surprisingly, the neural mechanisms involved in processing sound are similar to those in their bird cousins. For alligators, the evolutionary process solved the problem of hearing in two different media.

The same may be true for the sense of smell since alligators seem to be able to detect volatile airborne odors and water-soluble ones when submerged. Two nasal passages lead from openings on the tip of the snout through bones of the upper jaw to the back of the throat. Odors are detected by two types of olfactory sensory neurons that fill the nasal cavity. Combined, the two types may allow the alligator to smell above or below water. Once detected, incoming odors are passed along to the brain for interpretation. Well-developed olfactory regions in the brain indicate a good sense of smell. Alligators have two pairs of musk glands. One set, the gular glands, is located in longitudinal slits at the rear of the lower jaw. The glands can be everted to expose visible rosette-shaped structures. Paracloacal glands are located internally on each side of and emptying into the cloaca. Chemical analyses of the oily musk yield a complex brew of cholesterol, fatty acids, esters, and other organic compounds. During breeding displays of captive animals, a musky odor is often detected and likely plays an important part in the social behavior of alligators.

Alligators can see well above water also, especially on dark swamp nights when their nocturnal prey is active. The eyes are relatively close together and forward-looking to enable binocular vision that allows precise detection of objects of interest. Crocodilian pupils in bright light are shaped in the form of a vertical slit, but at night they expand to large globes for maximum efficiency. An artificial light aimed in the direction of alligators at night reveals the reflection of vivid silver to reddish-orange orbs. The cause of this eyeshine is a cluster of highly refractive crystals behind the retinas. Known as tapetum lucidum, these organs make the pupils appear to glow when struck by an outside light source. Animals with the brightest eyeshine usually have more rods and fewer cones in their retinas resulting in excellent night vision. Some reports indicate that alligators also have color vision. As violence is a way of life for alligators, they have acquired adaptations to protect the eyes. A bony brow shields the eye from above, and the eyeballs can be retracted in the sockets to some degree during feeding or fights. When submerged, a transparent eye cover (the nictitating membrane) slides sideways across the eye as a safeguard, and typical eyelids above and below can shut tightly when necessary.

Perhaps the most intriguing aspect of an alligator's umwelt lies in evidence that they can orient using stars. With the aid of celestial cues, alligators in outdoor pens were able to adjust to shoreline directions in a study that eliminated other possible signals. In essence, they possess a stellar compass the workings of which remain a mystery.

Dome pressure receptors, specialized organs to enable hearing and smelling above and below water, eyes designed to enhance survival in a perilous world, and the unexplained ability of celestial orientation—all combine to create the umwelt of an alligator.

Locomotion

Alligators move about in terrestrial and aquatic environments. However, with short legs and a streamlined body propelled by a long powerful tail, their locomotion is best adapted to water. When swimming, the tail moves from side to side in a wide sweeping motion. When captive alligators are placed in a type of aquatic treadmill and videoed, the wave patterns resemble that of fish. Thrust originates in the pelvic region. The arc of the tail sweep is constant, but the frequency of the sweep increases with speed of the animal. To minimize drag, the legs are held close to the body. The feet may occasionally be used to maneuver but play no important role in thrust. Hind feet are webbed but the front feet are not. Locomotion in water varies remarkably in degree from the almost imperceptible, slow disappearing act of a floating alligator to the violent thrashing of an alligator dismembering large prey.

On land, alligators use two distinct postures to move around. Both are considered types of plantigrade locomotion, which means they walk in a flat-footed manner. In the sprawl position the belly remains in contact with the ground and the legs propel the animal in a sliding fashion. This method is used most often for short-distance movement as when crawling out of the water to bask nearby. For me, one of the most impressive reptilian sights is that of a large alligator high-walking on land. The animal stands in the up position of a push-up as the body and all but the end of the tail are held off the ground. The high-walk posture is used most commonly for overland travel.

In rare instances alligators will gallop in a spectacular fast-forward version of the high walk. I once spotted a large alligator high-walking across a soybean field near the Tensas River NWR. When I ran toward him with a camera, he stopped and settled on his stomach before exploding in a gallop toward the river 75 yards away. I stood there agape oblivious of the camera as he crashed through the brush and disappeared over the high bank. A running alligator often appears to be moving faster than it actually is. For short distances they have been clocked at speeds up to 10 miles per hour—not fast enough to catch a horse in a sprint as the myth claims, but impressive nonetheless.

Locomotion of living crocodilians was once thought to be an intermediate stage between amphibians and lizards (considered sprawlers), and dinosaurs and mammals (erect, upright striders). Using alligators in laboratory settings, researchers have recently analyzed their movements in detail, proclaiming them unique, and concluding that crocodilians may not be a good example of the evolutionary transition from a sprawling to erect posture as once considered.

Temperature Regulation

Alligators, like all reptiles but unlike mammals and birds, do not have the ability to maintain a constant body temperature using internal physiological processes. They are "cold-blooded" or more correctly "ectothermic." In order to sustain a desired core body temperature in the range of 84° to 91°F, alligators have adapted behaviorally to exchange heat with their surroundings by moving to warmer or cooler places as necessary. To warm up, they bask on land and orient their bodies according to the amount of sunlight required. Lying at a right angle to the sun allows them to capture the maximum amount of radiant heat. Once warmed, they will face the sun to reduce exposure. If they begin to overheat they can open their mouth in a mouth-gaping posture for evaporative cooling or enter the water. The temperature of water is more stable than that of air because it warms and cools much slower. Accordingly, water tends to provide a thermal refuge during fluctuating air temperatures. Even in the water, alligators position themselves as needed for efficient thermoregulation by floating with their back out of the water or submerged except for the head.

Researchers in the Everglades implanted temperature monitoring data loggers in alligators and monitored them for a year. Their findings confirmed the idea that core body temperatures fluctuate with ambient (outside) temperatures throughout the year. In spring, alligators require more energy when they are most active as a result of breeding behavior and intense feeding after the winter dormant period. As detected in the experiment, alligators needed a constant, warmer core temperature then to produce this energy.

The least known aspect of alligator life history involves their behavior during the winter, especially in inland swamp habitat. In general, they retreat to dens during cold weather, but they do not hibernate. Instead, they brumate, a condition when the core temperature and other physiological processes decrease, but not to the extent that occurs in true hibernation. Other kinds of reptiles, including some snakes and turtles, also brumate. Alligators must surface to breathe when brumating and apparently move in and out of this state as the weather changes. They bask on warm winter days, but an alligator out of water on a very cold day is usually the sign of an illness. Their ability to slow down bodily functions allows them to survive cold weather only up to a point. Infrequently, extended periods of unusually cold weather, when water remains frozen for several consecutive days, occur in northern Louisiana and southern Arkansas. In the last 30 years I have observed alligator mortality within two weeks of almost all of these events. Usually the dead alligators were larger adults that floated to the surface. Larger individuals may have been more sensitive to cold or just more likely to be seen when they died. During a severe cold spell in the winter of 1983–84, thousands of alligators died in Louisiana, Texas, and Mississippi. The temperature dropped to 13°F in coastal marshes with an ice cover four inches thick for several days. Surveys showed that alligators of all sizes succumbed, and deaths continued for several weeks after the weather event. The extent of alligators' ability to regulate their body temperature limits where they can survive. As the climate warms in some areas, we might expect their range to expand.

4

To Build an Alligator

Growth and Size

Until the enlightenment provided by modern research, misinformation concerning the growth, size, and age of alligators was commonplace. An 1893 government document declared: "Alligators grow very slowly. At fifteen years of age they are only two feet long. A twelve-footer may be reasonably supposed to be seventy-five years old." Others claimed that alligators could live for 200 years and grow to lengths in excess of 20 feet. The truth is only slightly less remarkable.

Across the range of alligators, growth patterns are similar. As would be expected, those in areas with a more abundant food supply or more days with warm temperatures grow faster. Young are eight to nine inches long at hatching. Evidence from controlled experiments show that baby alligators in captivity hatched from eggs incubated at mid-range temperatures grow faster than those that were incubated at the ends of the temperature spectrum, at least until three years old. (Whether or not this holds true in wild populations is unknown.) Wild alligators do grow fastest during their first year of life. Measurements of thousands of known-age alligators allow mathematical predictions and show that growth slows over time. Both male and female juveniles grow at the same rate until three feet long, after which females grow much slower. Data reveal that males continue to grow at a fast pace for 20 years, reaching 8 feet 5 inches in length at age 10 and about 11½ feet at age 20 when they are growing 200 percent faster than females. Growth charts predict that at age 80 (an age that has never been verified in wild or captive alligators) the projected length of males is 13 feet 10 inches. Females average 6 feet 11 inches long at age 10 and 8 feet 5 inches at age 20. At age 45, females with a predicted length of 9 feet have nearly stopped growing. Another study from South Carolina found that male alligators there almost stopped growing in length when they reached 12 feet. Females exhibited the same pattern at 8½ feet. Both sexes did continue to grow in girth. Sexual maturity in both sexes is reached in 8 to 12 years at lengths of 6 to 7 feet. Females sometimes mature sooner than males. Although some north-south variation occurs, in most areas alligators stop feeding from October through March and thus do not grow during these months.

When humans ponder alligators, size matters. Consider the newsworthiness of a giant alligator, even in areas where the reptile is common. Throughout the recorded history of American alligators, accounts of colossal alligators are abundant and often accepted as truthful in spite of a lack of hard data or evidence to affirm such reports. It is as if there is a primal need to believe in the existence of potential human predators on a scale that exaggerates reality. Such ideas lurking around the edges of our domestic imaginations and roaring from the swamps of our psyche seem too much to give up. Make no mistake though; within the biological potential of the species, very large alligators have been scientifically documented and do still occur in the wild. The widespread, well-monitored harvests in recent years have yielded extraordinary individuals, but none that remotely approach the records in century-old claims.

Circumstances surrounding the report of what is usually considered the largest alligator on record merit a closer look. Edward Avery McIlhenny, called Ned by his family and friends, grew up on his father's plantation, which encompassed the ancient salt dome known as Avery Island on the Louisiana coast about 25 miles south of Lafayette. The family business included the production of Tabasco brand pepper sauce on this rare mound of elevated ground surrounded by thousands of acres of pristine wetlands. Natural productivity of virgin marshes in a near tropical climate is tremendous, and Avery Island featured a rich display of wildlife that included a cornucopia of estuarine species and clouds of wintering waterfowl. Alligators were likely as abundant here as anywhere on earth.

According to Ned, he and two assistants departed Avery Island on January 2, 1890, in a lugger and proceeded south through a maze of bayous to Vermilion Bay to hunt geese. Sailing southwest across the bay, they were becalmed at dusk near the mouth of a bayou that had been filled with silt by a hurricane a few years before. The shallow bayou, now cut off from the bay, led to Lake Cock several hundred yards inland. Ned, seeking game for supper, decided to walk the bank of the old stream with his shotgun. He shot two mallards just before dark, and when wading into the marsh to retrieve them, saw what he first thought was a partially submerged log. Approaching it, he discovered that it was a very large alligator that seemed to be addled in the cold air and water. He shot the alligator in the head, presumably with bird shot, and killed it. Upon close examination he discovered it to be the largest alligator he had ever seen. After spending the night on the boat, Ned and his companions went back the following morning to retrieve the carcass. Using a rope the three of them tried to pull the alligator through the marsh to the bank in order to skin it. The alligator was so large and the bottom of the marsh so boggy that they could only manage to move the animal a short distance. Ned then decided to abandon this effort and measure the alligator where it was. Using the barrel of his shotgun, which he knew to be 30 inches long, he measured the alligator three times. He then declared the total length to be 19 feet 2 inches.

It is a bit surprising to me that this record is still apparently accepted by many in the scientific community. If a similar claim were to occur today without corroborating evidence, it would almost certainly not pass muster. The fact that the record is based solely on Ned's word is problematic, even though he became a well-regarded naturalist and wrote a ground-breaking book entitled *The Alligator's Life History* in 1935. Material in his book is obviously based on long-term observations, experiments, measurements, and extensive notes. Ned was likely the most knowledgeable alligator expert in the country at one time. This and especially the fact that he became a prominent, wealthy businessman probably discouraged criticism of a record that otherwise might be questioned.

Issues pertinent to the record include Ned's age. He was 17 when he killed the large alligator. Even though he was unable to retrieve the specimen, it would not have been impractical to procure the head, a common tactic of contemporary naturalists. If one assumes that Ned's shotgun barrel was exactly 30 inches long, a standard length for such guns, the alligator would have been exactly 7.66 barrel lengths long—a dimension that in my experience of measuring large alligators in the marsh with steel tapes would be difficult to determine. Ned also mentions in his book two older alligator records greater than 18 feet from near Avery Island, though he did not personally observe these animals.

A look back into Ned's writings reveals some proven discrepancies. At one time he boasted of first introducing nutria to Louisiana after acquiring the original stock in Argentina. The McIlhenny Company website now states that this claim is untrue. The company historian has said of Ned that, "He was well-known on the island for his gift for spinning yarns. . . . I think he saw himself as an entertainer when relating his personal history. He took liberties in a good-natured way." Ned wrote the account of his record in his book published 33 years after the incident. With so little evidence, we will never know the truth of Ned's youthful encounter with the giant alligator.

Of course, there are other questionable claims of extremely large alligators. None, though, have been accepted like McIlhenny's. Researchers in Florida have investigated many such reports in that state. Using alligators that were carefully measured, they devised a mathematical model that allowed them to accurately predict the total length of an alligator based on the known length of its head. With this tool they were able to evaluate several old skulls and compare the declared length with the length predicted by the model. The skull of one "record" specimen was measured and the total length of the entire alligator was estimated to be 14 feet 10 inches. The early naturalist had claimed the animal was just over 16 feet long. Researchers found this trend to be the norm when investigating the old skulls.

It would seem reasonable to assume that the oldest alligators are the largest. The average and maximum life span of alligators in the wild is, however, a mystery. As it stands, there is no reliable way to determine the age of wild adult alligators. One promising yet unproven technique involves measuring their telomeres, which are DNA sequences on the end of each chromosome. Telomeres tend to shorten with age in most animals. Comparing telomere length of captive, known-age alligators may someday yield an index to age in wild animals.

Captive specimens have been reported to live more than 70 years. Life is much more precarious for wild alligators, and there is no evidence that they live longer than 50 years. Ned McIlhenny wrote that extremely large alligators mentioned in his book were likely "larger than normal, and were very old individuals, probably older and larger than alligators will ever again attain." His belief and that of some others was that intensifying hunting pressure would eliminate alligators before they could become old enough to reach maximum lengths. There are now examples that tend to discount this theory. In 1981 I organized a study on Lacassine NWR in southwestern Louisiana that involved capturing, measuring, and tagging alligators of all sizes. The 16,000 acre area of extremely productive freshwater marsh was an inviolate sanctuary that had been protected from alligator harvest since establishment of the refuge in 1937. There were large alligators present when the refuge was created, leading to the conclusion that very old individuals could exist when the research began. More than 3,000 alligators were caught during the first years of the study. A commercial harvest began there in 1983. Through the

2011 season more than 8,300 alligators were harvested. None of the alligators caught during the study or in the harvests exceeded 13 feet. Similar situations have occurred in other areas. A zoologist in Florida wrote in 1891, "The largest specimen I saw measured twelve feet in length; and none of the many hunters and natives of Florida I have met have seen any longer than thirteen feet." The bottom line seems to be that the maximum potential length of most male alligators falls between 14 and 15 feet, and that the probability of exceeding 15 feet is exceptionally low. That's a long way down the bayou from Ned's 19 feet 2 inch "record." Interestingly, McIlhenny reported that the largest female alligator he had ever seen measured 9 feet 1½ inches. Modern harvest records have yielded females considerably longer (for example, 10 feet 2 inches from Mississippi and Florida, and 10 feet 3 inches from Louisiana).

Detailed information has been gathered on hundreds of thousands of alligators harvested since intensive management of the species began in the early 1970s. As for verified records, the longest known alligators from each state are listed below along with weight, location, and date, if known. All were males.

Some observers report that alligators come in two general body types— long and lean, and shorter and stocky. As can be gleaned from the state records below, longer alligators are not necessarily heavier. Alligators put on weight throughout the growing season and are heaviest before winter denning; they then metabolize fat during the inactive period and emerge lighter in the spring. A 13 foot 10 inch male from Florida weighed 1,043 pounds in late summer. A record female, also from Florida, weighed 285 pounds.

Table 3. Largest Verified Alligators by State

State	Length	Weight (lbs.)	Location	Date
Alabama	14'2"	838	Alabama River	2011
Arkansas	13'1"	680	Arkansas Co.	2010
Florida	14'3.5"	654	Lake Washington	2010
Georgia	13'9"	692	Flint River	2010
Louisiana	14'1"	?	?	Largest in 250,000+ records, 1972–92
Mississippi	13'8"	?	Pearl River	2009
North Carolina	12'6"	475	South River	1981
South Carolina	13'6"	1,025	Lake Moultrie	2011
Texas	14'4"	est. 900	Jackson Co.	1998

Food Habits

Within the range of American alligators, if a creature walks, flies, swims, or crawls through alligator habitat it is a potential meal for this reptile. There are few exceptions. As such, alligators are considered predatory generalists in their dietary habits and opportunistic in their selection of foods, which means that basically, they eat whatever animals they can capture wherever they happen to live. Their prey varies according to the size of the alligator and to the abundance of prey species.

Prey size increases with the size of the alligator. Hatchlings soon begin feeding on insects and small fish. Juveniles eat a wider variety of insects, crawfish, crabs, frogs, and small fish. In one study apple snails composed more than half of the volume of stomachs of young, late summer alligators in the Everglades. As alligators grow they take more vertebrates (animals with backbones) including mammals, birds, other reptiles, and larger fish. Turtles, snakes, muskrats, nutria, and fish such as gar are important foods for adults in some areas.

Foods naturally vary among the different types of marsh habitats. For example, a study of large alligators in Louisiana marshes found mammals, par-

Alligators are opportunistic predators of most animals in their habitat. U.S. Fish and Wildlife Service, NCTC Image Library.

ticularly nutria, to be important foods in fresh, intermediate, and brackish marshes. However, in the more saline brackish marshes mammals were less important and crustaceans such as blue crabs, crawfish, and shrimp became more significant. In the Everglades mammals are not as important because mammal populations are low there also. In inland baldcypress swamp habitat, reptiles (mainly turtles) and amphibians were found most often in the stomachs of adult alligators, but mammals were most important in terms of weight and volume of samples.

An excellent example of the alligator's ability to adapt to faunal changes in its habitat can be found in the history of nutria introduction in coastal Louisiana. Food habit studies in the 1940s and 1950s showed that native muskrats were the most important mammal in the diet of adult alligators. Exotic nutria from South America were introduced into the region at several places during the 1930s and 1940s primarily as a new source of pelts for the fur industry. After gaining a toehold over a period of several years, nutria began to disrupt the marsh ecosystem by out-competing and displacing muskrats. By the early 1960s nutria were a significant part of alligator diets, and muskrats had declined in importance along with their overall population. As they have managed to do over the last 200 million years, the opportunistic alligators adapted. A new issue regarding the alligator/nutria connection surfaced in 2002 when Louisiana wildlife officials began a nutria removal program to reduce marsh loss caused by the herbivorous rodents. There was concern that exterminating nutria would have negative consequences on alligators since they were an important food source. Research showed that alligator stomachs from the region contained nutria remains at similar rates regardless of whether intensive nutria removal had occurred or not. The outcome of the project confirmed that nutria control is difficult even when humans and alligators work together.

Alligators also do not differentiate between other natural and non-native species. They will similarly eat dogs, cats, hogs, goats, and smaller cattle if available. One farmer abandoned an effort to raise hogs in Louisiana marshes because of intense alligator predation. Dogs are mentioned often as victims of alligator attacks and declared (unscientifically) by some to be favored food items. Today's news reports of family dogs lost to alligators have a long history of their own. A passage in an 1886 journal reads: "The alligator is daintily-choice in his food, preferring a dog to the piney-woods hog." An earlier writer in 1858 noted: "I had heard from good authority—the alligator hunter himself, who had often captured them by such a decoy—that these reptiles will follow a howling dog for miles through the forest, and that the old males especially are addicted to this habit."

Perhaps one of the most unusual examples of alligators' catholic diet was reported in an 1870 edition of the *Tallahassee Sentinel*. The menagerie associated with a traveling circus was attacked by alligators while attempting to ford a swamp between the capital and Quincy, Florida. Even though the circus owner had been forewarned of the numerous predators at the site, he attempted the crossing with one elephant, two camels, two dogs, and two horses. After a melee involving several large alligators, he exited the swamp with the elephant, one camel, and one horse.

Alligators will attack and consume humans; because such incidents are sensational, even if very rare, the phenomenon will be discussed in a separate section. They have also been reported feeding on carrion on occasion. On Georgia barrier islands alligators have been observed scavenging loggerhead sea turtle carcasses. Stomach analyses often yield plant materials, sticks, stones, and other inanimate objects. Some were manmade items like shotgun shell casings, fishing lures, rope, cans, light bulbs, nails, bolts, dog tags, bird bands, glass bottles, and bricks. Most were probably swallowed incidentally while capturing prey or accidentally as in mistaking stones for snails. Alligator eggshells also turn up occasionally in reports, likely a result of female attempts to free hatchlings from the nest.

Alligators are voracious cannibals. Only the largest, strongest individuals are immune. I have seen fairly large specimens horribly mutilated with missing legs and partial mandibles. In these cases the injuries were likely the result of territorial disputes. Had they not escaped, they could have been consumed by the victors. Most food habits studies detect cannibalism, and research designed specifically to determine levels of cannibalism have yielded varying results. One study estimated the annual mortality of juveniles to be 6 to 7 percent. Others have concluded the annual rate of cannibalism on an entire population to be anywhere from 2 to 50 percent. Factors that may influence the rate of cannibalism include habitat type, demographics of the alligator population, and availability of other natural foods. Larger alligators (greater than nine feet) tend to be most cannibalistic and prey on juveniles and small adults. Research suggests that in some areas each cannibalistic alligator consumes an average of just over two of its kind in a year. It is important to note that some degree of cannibalism is normal, but man-made stresses on a population may drive the natural control process askew. If, for example, fish were poisoned with chemicals in a Florida lake, alligator cannibalism might be expected to increase.

The following list reflects the diversity of alligator foods as depicted in stomach analyses of individuals of all ages in various habitats.

Table 4. Alligator Foods as Detected in Stomach Analyses of Alligators of All Ages in Various Habitats

Invertebrates

Apple snail	Fiddler crab	Other snail species
Blue crab	Giant water bug	Spiders
Clams	Grass shrimp	Water scorpion
Crawfish	Green June beetle	Whirligig beetle
Dragonflies	Mussels	White shrimp
Earthworm	Other crustacean species	
Eastern lubber grasshopper	Other insect species	

Amphibians

Bullfrog	Green tree frog	Two-toed amphiuma
Greater siren	Other frog species	

Birds

American coot	Great blue heron	Mottled duck
Anhinga	Great egret	Northern pintail
Barred owl	Green heron	Osprey
Blue jay	Hawk (unidentified)	Purple gallinule
Boat-tailed grackle	King rail	Purple martin
Cattle egret	Least bittern	Red-winged blackbird
Common grackle	Little blue heron	Ring-billed gull
Common moorhen	Louisiana heron	White ibis
Double-crested cormorant	Mallard	Wood duck

Fish

Alligator gar	Drum	Red shiner
Anchovy	Gizzard shad	Sailfin molly
Black bass	Golden shiner	Sheepshead minnow
Black crappie	Killifish	Silversides
Bluegill	Lake chubsucker	Spotted gar
Bowfin	Menhaden	Sunfish
Brown bullhead	Mosquitofish	Tarpon
Buffalo	Mullet	Tilapia
Carp	Needlefish	Warmouth
Catfish	Red-spotted sunfish	Yellow bullhead

Mammals

Armadillo	Goat	Opossum
Cattle	Hispid cotton rat	Rabbit
Cotton mouse	Hog	Raccoon
Dog	Mink	Rice rat
Domestic cat	Muskrat	Round-tailed muskrat
Eastern wood rat	Nutria	White-tailed deer

Reptiles

Alligator	Florida softshell turtle	Redbelly turtle
Brown water snake	Gopher tortoise	Red-eared turtle
Cooter	Loggerhead musk turtle	Ribbon snake
Cottonmouth	Mud snake	Stinkpot turtle
Crayfish snake	Other snake species	Yellow-bellied turtle

Alligator with catfish prey. Photo by Burg Ransom.

As top predators, alligators employ tactics of ambush and stealthy pursuit depending on the feeding opportunity. Unsuspecting prey is often captured when it approaches too closely to a motionless alligator with only its eyes and tip of nose above the water. Turtles and fish such as gar are often caught in this manner. If larger prey is detected swimming nearby or wading in shallow water, alligators will often submerge and suddenly appear beside their victim to seize it before escape is possible. I once observed an alligator attack a yearling deer as she waded into a small pond to drink. From a distance I was not aware of the alligator before the explosive assault. The doe was drinking calmly when the surface of the water detonated. On this occasion the struggle lasted only a few seconds until the deer managed to escape. A closer examination yielded a faint blood trail leading from the pond and a lurking six foot alligator. The deer likely survived because her attacker was not quite large enough to finish the job. In some cases alligators will overtly charge their quarry with no attempt of furtiveness. A swimming raccoon or a brood of flightless wood ducks can stimulate this behavior. Hatchlings often aggressively chase insects. Adult alligators have been observed lunging at nests in wading bird rookeries and capturing startled fledglings as they fell.

Regardless of how prey is approached, it is captured by a head-on thrust or by a side-swipe. This includes instances when alligators are feeding on "passive" foods such as snails or carrion. Small food items are swallowed whole. Hard-shelled animals like turtles are tossed about in the mouth and crushed. Large prey is shaken and torn into smaller pieces. Very large prey, like deer or livestock, elicit so-called death roll behavior also known as rotational feeding. In these cases, the alligator will grasp the prey and roll rapidly to tear away chunks of meat. The physics of the death roll is similar to that of the figure skater's tactic to speed up a spin by pulling her arms in close to the body. Alligators press their fore and hind legs against the body and forcefully throw their head and tail sideways to begin the maneuver. The resulting rotation is a savage display of power that easily dismembers the toughest prey.

Reproduction

Unless one is a single-celled amoeba, sex is complicated. In higher life forms, alligators included, reproduction requires two individuals, and therein rests the source of complexity. Genetic material from each must somehow be combined to form a new creature that will eventually be capable of the same gene-sharing scenario. If the route doesn't work out for a species, the cliff of extinction looms just around the bend. For alligators the process involves physiological preparation, courtship behavior, mating, nest-building, egg production and laying, incubation, and hatching of viable offspring. Nothing is more important in the life cycle of alligators.

The first indication of alligator courtship is audible. At close range it evokes emotions from times when humans had good reason to fear predators on a daily basis. The vocalizations of courting male alligators are termed "bellows." To me they are more like deep, rumbling roars with percussion-like overtones. The sound is often felt as much as heard, perhaps because males produce subaudible infrasonic vibrations just before the audible bellow. Researchers have further defined six distinct calls during courtship as bellows, bellow growl, low growl, deep grunt, cough, and hiss. The sight of a bellowing bull alligator is even more impressive. He points his head upward at a 45° angle, arches his tail, extends the throat pouch, and literally vibrates in the water sending concentric crests of small waves streaming outward down the length of his back. Bellowing is often accompanied by headslap/jawclap behavior in which the alligator simultaneously claps his jaws together and slaps the water with his head. Bellowing can occur sporadically anytime during the year but begins in earnest on warm April mornings soon after daylight. In areas with dense populations, alligators bellow in choruses. Females also vocalize and may initiate a chorus, but their calls are higher pitched with less resonance.

Olfaction also likely plays a critical but poorly understood role in courtship. Musk is emitted during bellowing and head slapping, and tends to disperse along the surface of the water. The function of the odors is speculative but may help to attract alligators of the opposite sex, define territories, or signal aggression to others of the same sex. When these primal sounds and odors emanate from a swamp or marsh in spring, it is a sure sign that mating season has begun.

Bull alligators bellow to attract mates and announce their presence to competing males. Unlike many animals with ritualistic combat behavior that tends toward bluff and rarely results in serious injury, alligators fight with terrible viciousness that often leads to horrific injuries or death for the less dominant individual. On a late March morning while fishing on Black Bayou Lake NWR in north Louisiana, I was fortunate to witness the power and intensity of one of these battles. A bull had been bellowing for several minutes from vegetation along the shore before I saw him swimming into open water directly toward another alligator. The second bull saw the first and made a beeline for him. Both were large alligators—one was about 12 feet long and the other 10 feet. There was no posturing; they closed quickly and erupted from the surface belly to belly, almost standing on their tails. Nearly a half ton of thrashing alligators crashed back into what moments before had been a serene lake draped in Spanish moss and steeped in birdsong. This encounter was brief, as the smaller bull had enough and fled the scene. All the while a 6 foot alligator watched the action from a few yards away. I assumed this one to be a female, and if she was searching for a dominant mate the choice had just been made much easier.

Although the annual reproductive cycle is driven by the release of certain hormones that results in the production of proteins for egg and sperm development, actual timing of the reproductive season varies across the range of alligators and is influenced by temperature on a year to year basis. For example, egg-laying begins in early June when average temperatures of March-May are warmest but not until late June or early July when average spring temperatures are low. On typical years the peak period of egg-laying is correlated with day length.

The onset of bellowing bulls causes a change in the daily movements of female alligators, especially in marsh habitat. Adult male alligators prefer deeper, open water on a year-round basis, while females tend to live in areas that are shallower and more heavily vegetated. During the courtship period females move to open water apparently in response to vocalizing bulls.

On average, courtship activity intensifies until early June when breeding takes place. Successful copulation can only occur if the male alligator is longer than the female. The male climbs onto the back of the female and uses his front legs to clasp her body behind her front legs. He then wraps his tail around and under her body until their vents align for penetration. His penis is inserted into her cloaca and sperm is released. He must be much longer than the female to make the necessary connection. Indeed, in captive situations females have been observed attacking and killing smaller males that attempted to breed with them. Females are capable of storing sperm in their oviducts until they ovulate and fertilization takes place, but not from one year to another.

An extensive study in coastal Louisiana revealed that alligators do not routinely exhibit fidelity to a single mate in a given breeding season. Genetic testing of more than 1,800 hatchlings from 114 nests over several years showed that 51 percent of the clutches were sired by more than one bull. This means that those females bred with more than one male, and sperm from their multiple partners was stored and eventually fertilized different eggs as they ovulated. The rate varied widely among years, and some females did mate exclusively with the same male for several years.

Not all adult females nest in a given year. The percentage of nesters may be density dependent, with high populations resulting in animals that may be stressed nutritionally, show high levels of antagonistic behavior among breeders, or lack suitable nest constructions sites. One report stated, "The number of mature females reproducing each year is rarely greater than 50 percent, but data on internest intervals [those years between nesting attempts] is lacking."

After breeding, females move back to their more secluded habitat in shallower waters. Here, nest-building begins. The actual nest sites vary within and among the various habitat types where alligators live. Nests are built in swamps and open marshes, on levees, and along the shoreline of natural and man-made lakes and ponds. Some females nest on the same site year after year. One study showed that even after a devastating hurricane that completely changed the appearance of a marsh, females returned to nest near their previous nest sites. Other research indicates that nests are often clumped in certain areas. No one knows why. Likewise, for inexplicable reasons, alligators in east Texas interior wetlands tend to select nest sites within 15 feet of a large tree and at least 450 feet from open water.

Actual nest construction occurs at night and takes about two weeks. A female begins by flattening all of the vegetation around the site. Then, using her mouth and feet, the alligator pulls up and tears away vegetation and begins to pile it all up in a mound. Sometimes several false starts are made with incomplete piles before she settles on a final product. Nest composition varies, as alligators use plants in the immediate area. Sawgrass, wiregrass, three-square, maidencane, and roseau cane are often used in marsh settings. In swamps, leaves, cattails, and other aquatic plants are common. I have found nests on the edge of uplands built from blackberry briars, pine straw, and honeysuckle vines. When completed, the nest often sits in the middle of an area stripped bare of vegetation for several feet around.

As the nest nears completion the female, using a hind leg, forms an egg cavity in the center of the nest. About a foot beneath the top surface, a hollow space several inches in diameter is molded to receive the eggs. When the eggs are laid more plant material is piled on to cover them, giving the entire structure the appearance of an inverted cone. Individual nests vary greatly in size but average about six feet in diameter and two feet high.

Egg-laying, also a nocturnal event, begins soon after construction of the egg cavity and is usually completed within an hour. Females are reported to lapse into a trance-like state while egg-laying and only become protective after covering the eggs. The number of eggs produced by a female is likely tied to her overall health. Alligators, like turtles, resorb calcium from their bones to form egg shells. The average clutch size of several hundred nests in a Louisiana marsh was 39 and ranged from 2 to 58. In north Florida a

Biologist with alligator egg at nest. Photo by Roger Tankesly, Mississippi Dept. of Wildlife, Fisheries, & Parks.

smaller sample yielded an average of 30 eggs. A Florida nest collected in 1925 contained 75 eggs; however, two females are known to lay eggs in the same nest on occasion. In Louisiana all egg-laying is usually completed in a 2 week period each year. In Florida egg-laying begins in early June, extends until the second week in July, and seems to be more protracted farther south. An environmental cue that helps to regulate the cycles of an organism's biological clock is called a zeitgeber. For alligators the zeitgeber for egg-laying is day length. Studies in Georgia and Louisiana showed peak egg-laying when day length was greatest.

Alligator eggs can vary slightly in size but are usually about the size of a goose egg. A fresh laid egg is pure white and translucent. Within 24 hours an opaque, chalky spot forms when the embryo attaches to the top of the egg. The band expands to the ends of the egg as incubation progresses. Electron microscopy and advanced x-ray techniques reveal five layers of the eggshell. Beginning on the outside, these include: (1) a dense outer layer made of calcite crystals; (2) a honeycomb layer of calcite crystals at right angles to those in the first layer; (3) a layer made of mostly organic particles; (4) a layer that attaches the shell to the egg membrane; and (5) the egg membrane. As incubation proceeds the egg becomes more porous.

Eggs incubate in the nest for 65 to 70 days before hatching. During this period the female rarely makes contact with the nest but often stays nearby. Contrary to common belief, the mother alligators do not always aggressively protect their nests. Egg predation by raccoons is common, and researchers report that antagonistic behavior toward data collectors is infrequent. It has been my experience that females at nests often hiss and sometimes make threatening advances but always retreat when pressed.

Wildfire during the breeding period, especially in marsh habitat, can be devastating to nests. More than half of all nests were damaged or destroyed by a summer wildfire on a Louisiana refuge in 1995. Those that survived were on wetter sites. At the opposite end of the environmental spectrum, flooding can also impact nesting success. Early season hurricanes or other severe storms that cause flooding of egg cavities can kill developing embryos. Tropi-

Female alligator guarding nest. Photo Tony Pernas, USDI National Park Service, Bugwood.org.

cal Storm Beryl destroyed all nests in one study area in 1988. A simulated experiment showed that eggs can survive two hours of flooding, but mortality increases from that point onward and is total at 48 hours of submersion.

Without a doubt, raccoons are the greatest natural threat to alligator nests. From North Carolina to Texas, raccoons eat alligator eggs. As many as half of all nests in some areas are destroyed by raccoons in some years. For reasons not clearly understood, raccoon predation can vary greatly from year to year in the same place, but researchers think that causal factors include the size of the raccoon population (which fluctuates, often as a result of disease), water levels (which impact nest accessibility), and the abundance of crawfish (which provides an alternate food source). In Louisiana most raccoon predation begins just as the eggs start to crack, and the pillagers will return to a nest repeatedly over several days until all the eggs are eaten. Female alligators do not routinely protect their nests from raccoons. An important point to remember is that predator/prey relationships in a natural setting are part of a checks and balances system in which both species benefit in the long term. A common misperception is that predators always regulate prey populations when in fact it is often the other way around. Harmful imbalance occurs when unnatural elements are introduced into otherwise functioning ecosystems (for example, the draining of a marsh or the pollution of a swamp). In spite of predation, fires, and floods, various studies have reported overall hatching success at rates from 35 to 68 percent. Therein rests the norm.

In the realm of biological wonders few surpass the method that determines the sex of American alligators (and many other reptiles). The process is called temperature-dependent sex determination (TSD) because the sex of young alligators is determined by the temperature of the eggs during incubation. Laboratory experiments have revealed that embryos develop into females at low and high average temperatures and males at intermediate temperatures. Constant incubation temperatures between 84° and 88°F will result in all females. Between 88° and 90°F, a mixture of sexes will develop. Incubation at 91–92° will produce all males, and mixed sexes will occur again at 93°. At 95°F, near the lethal maximum, all will hatch as females. The important temperatures for a natural nest are those in the egg chamber, which remain fairly constant, and not those outside the nest, which can fluctuate greatly. Several factors influence incubation temperatures in the nest. The location of the nest seems most influential. Nests built on levees or other elevated areas tend to be warmer, as are nests in dry marshes. Nests in wet marshes or shade are cooler. Within an individual nest that produces mixed-sex young, males often develop in one part of the nest cavity and females in another due to slight temperature differences. Across the landscape on any given year, alligators nest on a variety of sites thus ensuring the production of both male and female hatchlings.

Although it is now well-known that incubation temperature determines the sex of hatchlings, the biological processes behind the phenomenon remain a mystery. The two sexes are genetically identical with no distinguishing X or Y chromosomes, so how does temperature control such a critical aspect of life? Something outside of the genetic code is going on that causes a physical change in the organism. The paradigm is a target of the emerging biological field of epigenetics—the discipline that studies genetically identical organisms that are measurably different in some way. Scientists are working diligently to solve the alligator riddle, not for the sake of reptilian biology, but rather for the impacts on related human health issues such as cancer, rheumatoid arthritis, and schizophrenia.

The hatching process actually begins about a week before the young escape from the nest. Rapid growth of the baby alligator causes the egg to swell and crack lengthwise. It is at this time that raccoon predation often begins in earnest, probably because the eggs are easier to detect by smell. After a few days lateral cracks link to form a web of fissures around the egg. Young alligators have a temporary "egg tooth" on the end of their snout, actually just a hard protuberance that they use to puncture the inner membrane. By thrashing about they can free themselves from their protective womb. Before and after escaping the egg shell, baby alligators will call out with yelping croaks. If the mother is nearby she will often come to the nest and liberate the hatchlings by tearing off the top of the nest with her mouth. The young quickly scurry into the nearest water and into a world of imminent danger.

Root Hog, or Die

Movement

Folklore depicting a giant alligator using the same basking log year after year or inhabiting the same 'gator hole for decades might lead one to believe that these reptiles are sedentary. According to a number of studies involving thousands of alligators, nothing could be farther from the truth—at times some of the creatures are vagabonds. Scientists analyze alligator movements by marking animals that they later recapture or encounter during a harvest, or by the use of radiotelemetry. In this manner the home range of an individual can be estimated if enough encounters occur, as well as the distances traveled.

Wide-ranging movements probably do not begin until an alligator is about three years old. Newly hatched young immediately crawl to the water and congregate with the mother at her nearby den. The hatchlings, collectively referred to as a pod, stay in this vicinity and overwinter there with the mother. They emerge the following spring and begin the serious business of feeding on any animal they can subdue. Even though the female may breed and nest again, many of the yearlings stay together to spend their second winter close by in an ever-loosening group. By the third summer, juveniles have abandoned their siblings and some have contracted a case of wanderlust.

Research has revealed that immature alligators move greater distances than adults, and the distance that animals of all sizes move increases along with the time between captures or radiotelemetry readings. There seems to be no difference in the movements of male and female juveniles, but adult

Alligator tagged and outfitted for movement study. Photo by Ricky Flynt, Mississippi Dept. of Wildlife, Fisheries, & Parks.

males tend to move farther than adult females. Nesting females in particular limit their movements to the area of their nest. The work shows that alligators often travel a linear distance of 15 miles, but do occasionally range 25 miles or greater. One juvenile female moved 56 miles from her point of original capture in three years. An adult was observed swimming in the Gulf of Mexico more than 38 miles from the nearest land.

Perhaps the question that is left begging is why alligators move long distances in the first place. In areas subject to hurricanes and other intense storms, some movements may be involuntary. Storm surges have been known to carry alligators for miles. Hurricane Audrey that struck southwestern Louisiana in 1957 washed alligators at Rockefeller Wildlife Refuge up to 10 miles north. A juvenile alligator that was tagged in Cameron Parish, Louisiana, may hold the record for a storm-related movement. After Hurricane Ike came ashore near Galveston, Texas, in September 2008, the young alligator washed up on the beach at Padre Island National Seashore 304 miles from its release site. Alligators, especially juveniles, may disperse to avoid predation. Some may move in response to a declining food source

Mounted radio transmitter on adult male alligator. Photo by Ricky Flynt, Mississippi Dept. of Wildlife, Fisheries, & Parks.

or an especially rich one. In Florida some alligators were found to commute between their normal freshwater habitat and the brackish waters of estuaries where rivers meet the sea. Because of their low tolerance for saltwater, the visits were limited to short periods but were still long enough for them to exploit the lush food resources of the area. Droughts and floods can alter habitat and force alligators to seek new nesting and denning sites. In the case of adults observed moving across country in areas with sparse alligator populations, it is not unreasonable to assume they are seeking mates, but our current knowledge regarding motivations for travel often amount to informed speculation.

An animal's home range can be defined as that area it uses throughout the year in the course of its daily activities such as feeding, resting, breeding, and rearing of young. Using radiotelemetry, efforts have been made to determine the home range of alligators based on their known movements. Results of this type of study have varied tremendously (for example, less than two acres to more than 600 acres) probably because of the low number of animals actually monitored and the tendency for a long-distance movement to skew the results.

Most movement studies have been conducted in Louisiana marsh habitat and the Florida Everglades. Less is known about the movement of alligators in rivers, swamps such as Santee, Atchafalaya, and Okefenokee, and smaller inland wetlands. Research in east Texas interior wetlands suggests that the home range and movements of alligators there are restricted by the overall size of the wetlands and fluctuating water levels. Another study in the longleaf pine/wiregrass region of southwestern Georgia found that alligators used riverine habitat and seasonally flooded wetlands differently depending on their sex and size. Radiotelemetry showed that adult males stayed in the rivers year-round. Adult females and juveniles of both sexes were found in the rivers and in seasonal wetlands and often moved between these habitat types. Nesting in that area occurred only in the seasonal wetlands. This work revealed the necessity of protecting the entire matrix of inland wetlands in order to sustain healthy populations of alligators.

Mortality Factors

For most wild creatures the odds of a single young animal reaching adulthood are not good, and alligators are no exception. Florida biologists use the example of a nest with 35 eggs. On average, 15 live hatchlings will emerge but only 6 will survive to be a year old. Five of the yearlings will live until they are 4 feet long. Only 4 alligators out of the 35 egg clutch will reach maturity at about 6 feet in length.

The threats begin before the eggs hatch in the form of untimely fires or floods. Females, on occasion, will inadvertently crush their own eggs as they crawl over the nest. Raccoons are the most important egg predators throughout alligator range, and bears, opossums, otters, hogs, and crows also destroy nests in some areas. The results of a 13-year study in Okefenokee Swamp revealed that 69 percent of nests were destroyed during that time, and surveillance cameras showed that black bears were the primary predator. It has been suggested but not demonstrated that fire ants may kill emerging hatchlings. Young alligators are subject to predation by wading birds, especially great blue herons, big fish, otters, raccoons, and larger alligators. In unusual situations even adult alligators have been killed by other predators including bears and Florida panthers.

Adult alligators kill their cohorts in fights. Cannibalism is considered the most significant cause of natural mortality in juvenile and adult alligators, although the rate varies widely among populations. A study on a Florida lake indicated that cannibalism removes 6 to 7 percent of the juvenile population each year. Results from research in a Louisiana marsh showed that cannibalism was responsible for more than half of all hatchling mortality and 64 percent of mortality in alligators older than 11 months. Some alligators began exhibiting cannibalism when they reached 4½ feet in length.

Severe storms can cause outright deaths and debilitate survivors for long periods as shown by Hurricane Rita that struck southwest Louisiana in 2005.

Fire, natural and man-made, contributes to alligator mortality by destroying nests. U.S. Fish and Wildlife Service, NCTC Image Library.

Blood analysis of alligators after the hurricane yielded signs of severe stress caused by exposure to full-strength seawater when the tidal surge enveloped coastal wetlands.

Relatively little is known about the impacts of diseases and parasites on wild alligators, although they are not believed to be significant factors in general. *Aeromonas* bacterial infections have been implicated in the deaths of alligators in Florida and Arkansas. As a deadly, toxin-producing bacterium it occurs most often in reptiles being stressed by environmental changes such as exposure to contaminants. Certain types of blue-green algae produce liver and neurotoxins that may impact alligators. West Nile Virus (WNV) has been detected in wild populations of alligators in Florida and in captive animals in Florida, Georgia, and Louisiana. In 2002, one of the first outbreaks of WNV occurred at a Florida alligator farm and led to human health concerns. The disease causes skin lesions and neurological disorders in alligators and can be fatal. Lab work showed that infected alligators can pass the virus on to mosquitoes that bite them. The USDA gave preliminary approval for a WNV vaccine for captive alligators in 2011.

Most wild animals are infested with a variety of parasites that normally have little impact on an otherwise healthy host. (This, in spite of the fact that Herodotus is translated as declaring that insects enter the mouths of basking crocodiles and suck their blood until the animals die of exhaustion.) The parasitic fauna of alligators needs more study to understand their role in alligator life history. Preliminary work in Florida and South Carolina yielded tongue worms (named for their shape) that live in the respiratory tract, two species of roundworms, four species of flatworms, a leech, and a blood parasite. For some unknown reason, the blood parasite was more prevalent in female alligators from South Carolina than males from the same area. In unusual cases, alligators have been hosts to sea turtle barnacles.

The impacts of alligator harvests are discussed in chapter 8. Humans also unintentionally kill alligators in other ways. They are run over on highways and caught on trotlines and in the nets of fishermen. Trammel nets of commercial fishermen are especially lethal for ensnared alligators. Alligators are also often struck by boats, and along the Intracoastal Waterway and shallow bayous of south Louisiana, by my unofficial count, crew boats that service oil and gas facilities seem to be involved in a disproportionate number of alligator strikes, perhaps because of their hull design and propulsion system. In all of these cases, the impacts on alligator populations are likely negligible. However, in a small way they do contribute to the fact that very few alligators die naturally of old age.

Drowned alligator in commercial fisherman's hoop net. Photo by Paul Yakupzack, U.S. Fish and Wildlife Service.

The Keystone Species

"Keystone species" is an ecological term that can be defined as a plant or animal having a disproportionate impact on maintaining the well-being of an ecosystem. In some instances the removal of a keystone species can cause an ecosystem to cease to function in a natural way, such that populations of other plants and animals may crash, or conversely, the abnormal increase of some to the point that an imbalance occurs. The keystone metaphor refers to the single critical stone in an arch that because of its unique position supports the entire structure. Remove that stone and the arch collapses. Alligators throughout much of their range are considered a keystone species.

Why is the presence of alligators critical to the welfare of hundreds of other types of plants and animals in some areas? The answer lies in two aspects of their behavior. First, they are large creatures that routinely alter the physical characteristics of the environment. Like beavers, they are engineers of their own habitat and that of many other species. A low level reconnaissance flight over the Florida Everglades or Louisiana marshlands reveals their handiwork in the fashion of alligator holes, trails, and nest sites.

Especially in south Florida, ecological studies have teased out the connections. There, alligators use their snout, front legs, and tail to excavate ponds called alligator holes. Ranging in size from 8 to 50 feet in diameter, they are often dug down to the shallow limestone bedrock. Vegetation and sediment that is removed is piled along the edges of the hole, thus creating a drier site in addition to the wetter pond habitat. Before extensive alteration of Everglades hydrology, a natural wet and dry season occurred annually. Although the impacts of the cycles have now been altered, alligator holes and trails still provide critical refuge for fish, aquatic birds, reptiles, and amphibians during the dry periods.

Alligators, of course, benefit by having a concentrated food supply at these times. The only place that some fish survive extended drought to later repopulate wetlands is in alligator holes. Alligator ponds often harbor plant species found nowhere else in the vicinity. Drier sites of the pond edges and old nest mounds provide nesting habitat for egrets and herons. It has also been suggested that well-used trails, worn free of vegetation by alligators as they move between ponds and their nest sites, may serve as natural firebreaks to provide a different type of refuge for some plants and animals.

A unique example of linked life cycles is illustrated by turtles and snakes that lay their eggs in alligator nests. One study found that a third of investigated alligator nests also contained eggs of the Florida red-bellied turtle. Other turtles that use alligator nests include the Florida softshell turtle, mud turtle, and common musk turtle. In this symbiotic relationship the turtles usually lay their eggs in the nest before the alligator. Both species may benefit as the turtle gains a nesting site, especially during high-water conditions when other areas are flooded, and the alligator's food supply is potentially enhanced after the turtles hatch. In Louisiana, mud snakes sometimes lay their eggs in alligator nests. The species is one of the few snakes that exhibits maternal care during incubation (pythons do also). This alliance remains a mystery, as does that of the strange thermophilic (heat-loving) fungi that grows in warm alligator nests.

In addition to modifying their physical environment, a second trait that supports the keystone position of alligators pertains to their feeding behavior. As apex predators, alligators bask at the top of their food web. Detailed studies of other peak predators such as gray wolves, African lions, and killer whales reveal that these animals often help maintain their prey species at levels compatible with healthy ecosystems, (for example, wolves reduce the degree of overgrazing in the Yellowstone ecosystem by feeding on elk). Alligators play a similar role, but we are just beginning to recognize the complexity of such interspecific relationships.

Alligators and turtles often have symbiotic relationships. Photo by Burg Ransom.

Work in the Big Cypress National Preserve shows that the feeding habits of alligators help some of their prey species including invertebrates, frogs, rats, and mice. Most of the benefits occur as alligators consume turtles and snakes that would otherwise eat the other prey species. Quite likely, alligators perform more ecological tasks than we yet know. For example, they are known to eat large numbers of highly predacious gar, which themselves have few other predators. Even parasites, which are natural components in most ecosystems, are in the mix of connected strands. A certain type of flatworm found in alligators completes a critical stage of its life cycle in the bodies of spotted gar.

In Florida wetlands and many coastal marshes the keystone rank of alligators is unquestionable. Their mere presence in healthy populations enhances biological diversity. The ongoing restoration of the Everglades is mind-boggling in complexity and magnitude, but its success is monitored in the well-being of resident alligators. In their special roles as coal mine canaries of

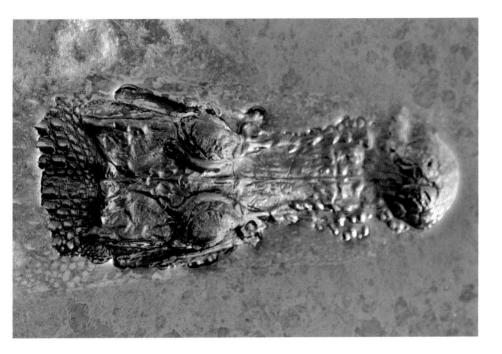

wetland vigor, alligator health will be an issue in the even greater challenges of restoring Louisiana's coast. The weight of the alligator keystone in habitats such as inland swamps and riverine systems is not as precisely known. It may be slightly less, or the arch may be so altered by humans as to make a true measurement impossible.

Part II

The Human Factor

See ya later, alligator.

First Encounters

Alligators and the First People

The first people arrived in the southern half of North America soon after the last ice age ended about 13,000 years ago. Climate change was ongoing, although at that time it was not caused by human behavior. In general, the climate was undergoing a slow warming trend after the maximum extension of the Wisconsin glacial period peaked 18,000 years ago. Sea levels were also rising back up over the continental shelf after having fallen more than 400 feet when water was tied up in the glaciers. Alligators as we know them were present, but their range was likely more restricted to southern coastal areas because of the cooler, drier climate and less abundant inland wetlands. The first humans here also encountered animals that were soon to vanish from the face of the earth. Saber-toothed cats, woolly mammoths, and mastodons became extinct about 10,000 years ago.

As the climate warmed, silt-laden melt waters from the northern glaciers filled the scoured valley of the lower Mississippi River to form a giant floodplain 500 miles long and nearly 200 miles wide. Meanderings of lesser streams across this Mississippi Delta created alligator habitat in the form of diverse wetlands. Sediments that reached the mouth of the Mississippi River were swept westward by gulf currents to create the most productive marshlands in North America along the Louisiana coast. On a smaller scale similar geologic processes were creating modern alligator habitat along the east coast. Florida's land-base was shrinking as the rising sea drowned its flat topography and increasing rainfall filled the pock-marked, limestone bedrock with wetlands as varied as small, cypress-draped springs and the vast Everglades. As environmental conditions gelled into habitats with the necessary parameters of food, water, and moderate annual temperatures, alligators expanded their range accordingly—and humans were not far behind.

Paleo-Indians, as these first people are termed, probably lived in small mobile bands as hunters, scavengers, and gatherers of various resources throughout the year. They left behind a trail of high quality, stone projectile points used to kill the now extinct megafauna. There is no evidence that they hunted alligators, but they lived among them, they certainly had the capability, and there is no reason to think that they would not have exploited the reptiles on occasion as did their descendants to come. Beginning about 9,000 years ago and after the extinction of the giant mammals, the Archaic Culture developed in which humans became more sedentary by establishing semipermanent seasonal camps and increasing the variety of their diets. By 3,000 years ago sea level had stabilized at its modern position and wetlands rich in natural resources were abundant across the Southeast. The stage was set for permanent human habitation of alligator habitat throughout its range. As variations of ensuing cultures (such as Woodland and Mississippian) continued to evolve, Native American populations expanded to fill the niches.

It is in this time period that modern archaeology begins to document human/alligator interactions, and the sites range from present-day Texas to North Carolina. On Sapelo Island, Georgia, remains of alligators were found in the complex of man-made shell rings dated to 4180 BP ("before present," defined as 1950 in relation to radiocarbon dating). At the 3,500 year old Poverty Point site in northeastern Louisiana, alligators were consumed in the shadow of a giant, enigmatic bird mound. Excavation of a Late Woodland burial mound (3200–1000 BP) near Sarasota, Florida, yielded 429 human skeletons, 4 dogs, and an alligator. The alligator was ceremonially interred with grave goods and appears to be the last burial in the mound. People of the Tchefuncte Culture (3000–1200 BP) lived along coastal areas from eastern Texas to Florida and inland to southern Arkansas. They were among the first in the region to routinely make pottery. These vessels were used to prepare the cornucopia of foods including alligator that were gleaned from surrounding wetlands. On a barrier island near Fort Lauderdale, Florida, scientists excavated a Tequesta Indian village site dated 1100 to 1500 and discovered alligator scutes thought to be the remnants of meals.

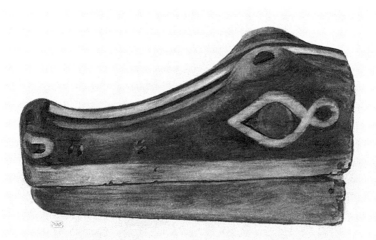

Watercolor of carved alligator head attributed to the Calusa tribe and excavated at Key Marco, Florida. National Anthropological Archives, Smithsonian Institution, INV 08517000.

A spectacular carved alligator head attributed to the Calusa tribe was also unearthed in Florida. Alligator teeth were exotic trade items valued by Native Americans living outside the range of the reptiles. Teeth believed to have Florida origins (based on other artifacts found nearby, like seashells) have been discovered in Hopewell Culture ceremonial sites in Ohio. Alligator effigy mounds, earthworks shaped like an alligator, have been reported from the Gulf Coast. One made of marine shells in the southeast corner of Grand Lake, Louisiana, is reported to have been 600 feet long but was suffering from significant erosion by 1935. A famous mound near Granville, Ohio, called the Alligator Effigy Mound, however, almost certainly does not represent an alligator.

Beginning in the sixteenth century, European explorers provided the first written documentation of Native American/alligator interactions. Jacques Le Moyne was a member of an ill-fated French expedition in 1564 to colonize north Florida. In the vicinity of present-day Jacksonville they built a fort and encountered local Indians. Passages of Le Moyne's narrative on the subject of alligators, which he called crocodiles, are worthy of note:

At a set time every year they [local Native Americans] gather in all sorts of wild animals, fish, and even crocodiles. . . . Their way of attacking crocodiles is as follows: They put up, near a river, a little hut full of

Le Moyne's sixteenth-century image of Native Americans hunting alligators in Florida. Courtesy of the John Carter Brown Library at Brown University.

cracks and holes, and in this they station a watchman, so that he can see the crocodiles, and hear them, a good way off; for, when driven by hunger, they come out of the rivers, and crawl about on the islands after prey, and, if they find none, they make such a frightful noise that it can be heard for half a mile. Then the watchman calls the rest of the watch, who are in readiness; and, taking a portion, ten or twelve feet long, of the stem of a tree, they go out to find the monster, who is crawling along with his mouth wide open, all ready to catch one of them if he can; and with the greatest quickness they push the pole, small end first, as deep as possible down his throat, so that the roughness and irregularity of the bark may hold it from being got out again. Then they turn the crocodile over on his back, and with clubs and arrows pound and pierce his belly, which is softer; for his back, especially if he is an old one, is impenetrable, being protected by hard scales. This is their way of hunting crocodiles. . . . In order to keep these animals longer, they are in the habit of preparing them as follows: They set up in the earth four stout forked stakes; and on these they lay others, so as to form a sort of grating. On this they lay their game, and then build a fire underneath, so as to harden them in the smoke. In this process they use a great deal of care to have the drying perfectly performed, to prevent the meat from spoiling.

A Spaniard familiar with south Florida in 1575, Hernando D'Escalante Fontaneda, reported that Indians in that region also ate alligators. Robert de La Salle on his noted expedition down the Mississippi River in 1682 observed Natchez Indians eating smoked alligator meat. European explorers in coastal Texas witnessed Atakapa and Karankawa Indians smeared with alligator fat to repel mosquitoes. These tribes were also reported to kill alligators by spearing them in the eye. Others stated that Native Americans in Florida considered alligator teeth potent medicine, especially against snakebite (possibly because they saw alligators eating snakes without harm), and often wore an alligator tooth necklace. Creek Indians performed sacred alligator dances.

The folklore of southeastern Native Americans was rife with alligator characters. Often depicted as dangerous dimwits in roles opposite clever creatures like rabbits, they received the blunt end of moral lessons. In other instances alligator spirits were summoned to perform evil deeds. Less frequently, they were depicted as benevolent protagonists who aided deserving people.

By the mid-nineteenth century Native American cultures across the Southeast were collapsing under the weight of European-wrought devastation in the shape of disease, displacement, and outright genocide. From this point forward their association with alligators would resemble that of their conquerors. The unique multimillennial relationship between Native Americans and American alligators survives in scraps of folklore as described by Bill Day of Louisiana's Tunica-Biloxi tribe: "It is said that in the beginning there were two great alligators, one red, one blue. They moved aside and allowed the Tunica to pass through into this world. There is red mud here and a blue mud, so who knows? In every myth there is an ounce of truth." It also lingers in the traditional alligator dance of the Caddo and Seminole, in the Alligator Clans of Seminole and Creek tribes, and in each language's word for *alligator*:

chinchuba:	Choctaw
tsu-la-s-gi:	Cherokee
allapatta:	Seminole
koo-hooh:	Caddo
tamahka:	Tunica
akshoti:	Ofo
nuxwoti:	Biloxi

The New People—Early Years

While conducting research for various writing projects, I have read thousands of old letters, diaries, and journals to glean first-person accounts of natural history. At least until recent times, most people seem to have had an affinity for nature and often wrote about interesting flora and fauna, especially during travels to unfamiliar regions. With this in mind, one of the most unexpected surprises that surfaced in researching this book pertains to the absence of information regarding alligators encountered on two major European expeditions to the New World. Álvar Núñez Cabeza de Vaca was a member of the Narváez expedition to Florida in 1527. He spent several years wandering about in prime alligator habitat from Florida to Texas. When he returned to Spain he wrote extensively about his journeys, often in rich detail, but never once did he mention alligators. The second example involves the De Soto expedition of 1539–43 and its destructive ramblings throughout the Southeast. Three different participants in the travels penned thorough chronicles, often mentioned plants and animals, but never commented on alligators. (One writer did allude to the American crocodile in Cuba.) We are fortunate that many who followed did not hesitate to document their alligator observations.

As mentioned in the previous section, artist Jacques LeMoyne encountered alligators in north Florida in 1564. Barely escaping a Spanish massacre, he survived to sail back to France where he drew pictures of alligators, perhaps the first European documentation of the reptiles. In 1699, Sieur de Sauvole, a leader in the French colony established near present-day Biloxi on the Gulf Coast wrote: "There is such a great supply of alligators that one sees them at every moment. But we have not the occasion to complain at the present; we have killed several at the foot of the fort; they don't come back so frequently any more." Antoine-Simon Le Page du Pratz arrived in Louisiana from France in 1718. As a naturalist and historian, he is remembered for his *Histoire de la Louisiane*, an account of his years in the lower Mississippi Valley in the first half of the eighteenth century. In this book he writes of alligators, which he called crocodiles, and in addition to repeating common misconceptions, made some surprisingly accurate remarks concerning their natural history.

Early European encounters with alligators often resulted in exaggerated reports especially as it pertains to the reptile's ferocity and size. Louis Judice, a French official and naturalist writing about a 1772 expedition southwest of New Orleans, stated: "But upon our arrival there, we encountered an alligator of prodigious size which lunged at us, forcing us to save ourselves on land. We estimated it to be thirty-five feet long."

William Bartram was an early American naturalist who wrote vivid accounts of his experiences with alligators in his widely read *Travels through North and South Carolina, Georgia, East* and *West Florida, the Cherokee Country, etc.* first published in 1791. The book made him famous, but unfortunately his narrative of alligator behavior also seems embellished with poetic license and littered with misinterpretation and/or exaggeration. Even today, two opposing camps argue the accuracy and thus scientific value of Bartram's work. In my opinion his writings of alligators have merit and should be considered in light of the science of his day. The following are examples of controversial passages and refer to incidents along the St. Johns River in 1774.

As I passed by Battle lagoon, I began to tremble and keep a good lookout; when suddenly a huge alligator rushed out of the reeds, and with a tremendous roar came up, and darted as swift as an arrow under my boat, emerging upright on my lee quarter, with open jaws, and belching water and smoke that fell upon me like rain in a hurricane. . . . the monster came up with the usual roar and menaces, and passed close by the side of my boat, when I could distinctly see a young brood of alligators, to the number of one hundred or more, following after her in a long train. . . . Behold him rushing forth from the flags and reeds. His enormous body swells. His plaited tail brandished high, floats upon the lake. The waters like a cataract descend from his opening jaws. Clouds of smoke issue from his dilated nostrils. The earth trembles with his thunder. . . . Only the upper jaw moves, which they raise almost perpendicular, so as to form a straight angle with the lower one. . . . The nests or hillocks are of the form of an obtuse cone, four feet high and four or five feet in diameter at their bases: they are constructed of mud, grass, and herbiage. At first they lay a floor of this kind of tempered mortar on the ground, upon which they deposit a layer of eggs, and upon this a stratum of mortar seven or eight inches in thickness, and then another layer of eggs, and in this manner one stratum upon another, nearly to the top. I believe they commonly lay from one to two hundred eggs in a nest.

John James Audubon is renowned as the most famous American wildlife artist. His painting skills were fueled by an acute power of observation that is reflected in the details of his bird, mammal, and plant images. Always the consummate naturalist, he caught baby alligators to observe them up close and reported shooting a large alligator in the St. Johns River in 1832 in order to accurately sketch its head. In a letter to a natural history society he gives an early nineteenth century account of the natural history of alligators. A small part of the lengthy treatise follows:

> In Louisiana, all lagoons, bayous, creeks, ponds, lakes, and rivers, are well stocked with them,—they are found wherever there is sufficient quantity of water to hide them or to furnish them with food, and they continue thus, in great numbers, as high as the mouth of the Arkansas River, extending east to North Carolina, and as far west as I have penetrated. On the Red River, before it was navigated by steam-vessels, they were so extremely abundant, that, to see hundreds at a sight along the shore, or on the immense rafts of floating or stranded timber, was quite a common occurrence, the smaller on the backs of the larger, groaning and uttering their bellowing noise, like thousands of irritated bulls about to meet in fight, but all so careless of man, that unless shot at, or positively disturbed, they remained motionless, suffering boats or canoes to pass within a few yards of them, without noticing them in the least. The shores are yet trampled by them, in such a manner that their large tracks are seen as plentiful as those of sheep in a fold. It was on that river, particularly, that thousands of the largest size were killed, when the mania of having either shoes, boots, or saddlebags, made of their hides, lasted. It had become an article of trade, and many of the squatters, and strolling Indians, followed, for a time, no other business.

Audubon painted his classic whooping crane in New Orleans in 1821. Two baby alligators were added to the foreground the following year.

Americans have been involved in the commercial trade of alligators since at least 1800. Audubon reported in 1827 that "as water levels receded alligators congregate into the deepest hole in vast numbers and to this day, in such places are shot for the sake of their oil, now used for greasing the machinery of steam engines and cotton mills." An 1846 journal portends the alligators' future: "Your alligators are looking up. They have been considered dull, stupid wretches; but are now discovered to have a world of light in them, when properly extracted and kindled: in a word, they are to be killed for their oil. We have almost used up whales, and shall now begin to burn the midnight alligator." Thousands of alligators were also killed for their skins in this period but the popularity of alligator leather waned when according to Audubon "the discovery that the skins were not sufficiently firm and close grained to prevent water passage put a stop to the general destruction of alligators, the effect of which had already become very apparent." One researcher notes that the demand for skins on a large scale resurfaced in 1855 when leatherworkers in Paris sought hides for shoes, boots, and saddlebags.

Recent archaeological work focusing on the tragic institution of antebellum slavery shows ties between those in bondage and alligators. Excavations at Louisiana plantation sites yielded evidence that slaves often used locally available wild resources including alligators for food. A South Carolina slave said of alligators, "eat every part but don't eat the head and feet. Eat body part and tail." Artifacts unearthed in slave cabin sites on Ossabaw Island, Georgia, such as alligator teeth and raccoon bones, items often found in a mojo bag, may reflect the inhabitants' spiritual beliefs. A Texas slave woman noted that children wore alligator teeth along with rattlesnake rattles around their necks to alleviate teething pain. Solomon Northup was enslaved near the Red River in Louisiana when he made a desperate attempt to escape. Pursued by hounds, he ran to the nearby swamps for haven and later wrote: "I saw also many alligators, great and small, lying in the water, or on pieces of floodwood. The noise I made usually startled them, when they moved off and plunged into the deepest places. Sometimes, however, I would come directly upon a monster before observing it. In such cases, I would start back, run a short way round, and in that manner shun them. Straight forward, they will run a short distance rapidly, but do not possess the power of turning. In a crooked race, there is no difficulty in evading them."

The American Civil War (1861–65) exposed people to alligators in new and different ways. A leather shortage in the South roused a fresh demand for alligator hides, and thousands of the animals were killed for shoes, boots, and saddles for Confederate troops. Reflecting the general attitudes of the era, many Civil War participants considered alligators in the same category as bears, panthers, and wolves—vermin with no intrinsic beneficial values other than as targets for rifle practice. When mention of this species was made in their writings, this belief prevailed. Perhaps because many Southerners were familiar with alligators, the species is more commonly mentioned in Union accounts, a result of its novelty. The following passages from across the Southeast underscore the belief that just the sight of an alligator was worth noting in a letter or diary.

Private John Westervelt, 1st New York Volunteer Engineer Corps, near the mouth of the St. Johns River, Florida, on February 27, 1864: "The banks of the river are mostly low and marshy. Occasionally we saw an alligator basking his horny hide in the sun."

Private Theodore F. Upson, 100th Indiana Infantry Volunteers, near Goldsboro, North Carolina, on March 27, 1865: "There are a great many small alligators and once in a while quite a large one in the pond above the mill. The boys have shot several. There was one that has kept well away but has been seen at times. I got on top of the mill to day and he showed up a long shot away. I raised the sights on my rifle and was fortunate enough to kill him. When the boys got him he measured 7 feet in length. The citizens and Darkies here think I am a wonderful shot."

John S. Jackman, 9th Kentucky Infantry, aboard the steamboat *Waverly* on the Alabama River on September 29, 1862: "The boys amused themselves by shooting at aligators lying out on the sand-bars and banks, sunning themselves."

Corporal Rufus Kinsley, 8th Vermont Regiment, near Des Allemands, Louisiana, on June 3, 1862: "We killed four alligators on the way. I tried my rifle on two of them; put a ball in the right eye of each. One of them was thirteen feet long. We ate two of them for supper. Found the flesh, when boiled, more like a chicken's breast than any thing else."

Private Galutia York, 114th New York Volunteer Infantry, in a letter to his parents on February 1, 1863, from a ship near the mouth of the Mississippi River: "I saw an alligator day before yester day I should think he was 12 feet long they are thick down hear."

Captain Charles B. Haydon, 2nd Michigan Infantry, near Vicksburg, Mississippi, on June 27, 1863: "The country is not so bad after all as I was at first led to believe. There are not so many snakes or other infernal machines as was represented. The alligators eat some soldiers but if the soldiers would keep out of the river they would not be eaten." [Then as now, the irrational fear of alligators led to wildly inaccurate speculation.]

One Civil War anecdote pertaining to alligators was transcribed during an interview with a former Mississippi slave. Mr. Hamp Kennedy relates: "One time I 'member when Aunt Charity an' Winnie McInnis . . . tried to swim some of our hosses cross de riber to save 'em from de [Yankee] soljers an' dey rode 'cross in a little boat. Well, when de hosses got in de middle of de water, up comes a 'gator, grabs one hoss by de ear, an' we ain't neber seed him no mo."

During the Civil War, a vessel named the USS *Alligator* was the first submarine in the U.S. Navy. She was launched in 1862 and sank in 1863 after an insignificant career.

7

Near Fatal Attraction

Beyond Sustainability

After a brief respite following the war, alligator leather goods became fashionable again around 1870. From that point forward until today there has been a continuous market for the hides. Also beginning at that time and in accord with other American wildlife such as bison, grey wolves, and grizzly bears, alligator populations throughout its range began an accelerating decline that many biologists thought ended abruptly on the cliff edge of extinction. In retrospect, the cliff edge for alligators may have been a bit farther out than feared. Although alligators in many areas were completely eliminated, with protection other populations thought to be depleted recovered at rates faster than the anticipated natural population growth—indicating viable numbers remained in some places.

Nevertheless, the cause of the dramatic declines was reckless overexploitation. The number of alligators killed has been guessed at by various investigators based on their assessments of other estimates. A Florida researcher said that at least 2,500,000 alligators were killed in that state from 1880 to 1893. Another in Louisiana stated that up to 3,500,000 were killed there between 1880 and 1933. In Georgia, 10,000 skins were amassed annually between 1922 and 1926. Taxes collected on alligator skins by the Louisiana Wildlife and Fisheries Commission reveal that 314,404 skins were sold in the state from 1939 to 1955. The grand total of alligators killed across the Southeast between 1870 and 1960 has been estimated at 10,000,000. It is a reasonable guess.

The skin trade drove exploitation, but the cumulative demands on alligator populations for other purposes contributed to overall decline. The U.S. Bureau of Fisheries reported in 1884: "The ivory is obtained from the teeth. These are carved into a variety of forms, such as whistles, buttons, and cane-handles, and also sold as jewelry. This industry is carried on principally in Florida. Alligator oil, which is extracted from the fat of the animal, has been recommended for the cure of quite a variety of diseases. The musk of the alligator is obtained from glands situated in the lower jaw. It is not of the best quality, but serves as the basis of certain perfumes."

Another writer describes a late nineteenth-century visit to a business in New Orleans on Charters Street: "This establishment makes a speciality of supplying tourists from colder climates with living souvenirs of the district in the shape of little black alligators that have just shaken off their shells." [When a barefoot Cajun wearing a palmetto hat came into the business with a grass sack of baby alligators over his shoulder,] "They were disposed of to the alligator merchant at 5 cents apiece, to be retailed by him subsequently at 25 cents." Untold thousands of small alligators were carried home to northern states as pets by tourists visiting the South. Most perished for lack of proper care or succumbed when released into the hostile environment.

Forty-three different people from the surrounding area donated their pet alligators to the Zoological Society of Philadelphia in 1918 alone. Many others were shipped from the swamps to supply the demands of countless menageries that cropped up around the country. A 1900 *New York Post* article describes a reporter's experience with "a hunter of menagerie stock" on Florida's Caloosahatchee River. An 1886 journal confirms the exploitation: "By thousands and thousands the guileless alligator of tender years has been ruthless torn from the maternal breast and sent adrift upon the frozen North; hence, the alligator in a menagerie is as familiar as the ubiquitous monkey."

As was the case with bison, passenger pigeons, and wading birds decimated for the millinery trade, people were aware of the ongoing, unsustainable slaughter. The *Chicago Tribune* in 1889 noted: "Alligators have learned to avoid mankind and steamboats. Repeated trips can now be made the entire length of the St. John's River without seeing a single specimen. The reptiles have been hunted down until they conceal themselves at the first sound of a boat or the sight of any human kind." A researcher for the Smithsonian Institution (who had just collected 1,000 alligator eggs in Florida) stated:

In the summer of 1906 the Okefenokee was again visited; this time the swamp was penetrated to its centre, and nearly one hundred alligators were killed by the three hunters with whom I was traveling. It is this vigorous hunting, done chiefly at night, with a bull's eye lantern and shot gun, that has so diminished the numbers of alligators that where, twenty years ago, hundreds could be seen, to-day scarcely one may be found. It seems a very wanton destruction of life to kill so many of these large animals, especially when it is remembered that a large alligator hide is worth to the hunter only about $1.50. Just how soon the alligator is likely to be exterminated in our southern states it is impossible to say.

The decline was obviously noticeable to those whose livelihoods were tied to alligators, such as shoe manufacturers. Their trade magazine, the *Shoe Workers' Journal*, claimed in 1906: "The alligator is disappearing. The great American public, hungry for alligator slippers, alligator bags and other novelties of the rough, scaly skins, is causing the demise of thousands of alligators weekly."

In 1913, *Forest and Stream* magazine cited a special obstacle for alligators compared to other species in decline:

If the feather-bedecked hat implies the destruction of birds of plumage, the alligator skin hand bag means the passing of the 'gator; and with all his ugliness the Florida alligator bids fair to follow the Florida plume birds with all their beauty into the limbo of wild species destroyed for commercial purposes. One unfortunate feature of the case is that the alligator has no friends. He is universally regarded as an ugly customer. His ways are the reverse of winning. No Audubon Society espouses his cause. The sentiments evoked in behalf of the feathered singers in the trees has no regard for the alligator bellowing the swamp. The alligator must go.

Pressure on alligator populations was amplified by a number of factors. The high market demand generated more hunters and more hunter efforts in the swamps and marshes. Just before and during the Great Depression wide-scale trapping of furbearers began in alligator habitat across the Southeast, and a lifestyle developed to harvest wetland resources throughout the year. People trapped furbearers during the winter, gleaned fish and seafood throughout the year, and pursued alligators opportunistically. Technological advances, most notably gasoline boat engines, made it possible to access remote areas. Before powered skiffs, an alligator hunter's range was limited to the distance he could paddle in a day or two. In the vast marshes of Louisiana the development of oil and natural gas deposits quickened the rate of exploitation. Mineral exploration crews who systematically covered every acre of coastal marsh and routinely killed and skinned alligators in their daily work have been implicated as major players in the decline of that state's alligators. Oil and gas wells required the construction of access canals that attracted easily harvested alligators. At the same time throughout alligator range the reptile's habitat was besieged by industries and agriculture bent on "reclaiming" wetlands that were thought to have no value. In 1931 the Louisiana Department of Conservation declared, "From toll taken of this saurian during the past half dozen years, particularly during the drouth summers of 1924 and 1925, it seems at this writing that this giant and characteristic reptile is doomed to disappear from our fauna—and within the next few years."

When it comes to putting hard numbers on wildlife populations, biologists are well aware of potholes and roadblocks that hinder the task of such census-taking. The same is true when postulating cause and effect scenarios. Regarding alligators, one thing is certain though—healthy populations numbering in the millions thrived before intense commercial hunting began in the late nineteenth century, and this activity aggravated later by habitat degradation drastically reduced wild alligators throughout their range.

A striking view of alligator population decline is reflected in the following table that depicts the number of hides harvested in Florida each year from 1929 to 1943. The prices listed are those paid for a number one grade seven-foot hide. As alligators became less common, the price of hides rose and hunting pressure intensified. The temporary increase in harvested skins in 1935 and 1936 is thought to be during a time when Florida hunters expanded their efforts into Georgia.

By 1960 the survival of the alligator as a viable component of wetland ecosystems anywhere in its historical range was in doubt. Already over much of that region, the species had vanished. The only secure populations were found on refuges providing inviolate sanctuary. Prominent alligator scientists in Louisiana, once the heartland of millions of the reptiles, declared that "something had to be done, or the animal would soon be nearing extinction" and "not even a basic breeding population was present in many of the state's swamps and marsh areas."

Efforts to address the crisis were slow in coming and initially halfhearted. Political obstacles were a hindrance and included the twisted logic that pro-

Table 5. Reported Hides and Prices of Alligators Harvested in Florida, 1929–43

Year	Number of Hides	Dollar Price per Hide
1929	190,000	1.50
1930	188,000	2.50
1931	150,000	2.75
1932	145,000	2.75
1933	130,000	2.75
1934	120,000	3.00
1935	162,000	3.00
1936	150,000	3.00
1937	130,000	4.00
1938	110,000	4.00
1939	80,000	5.25
1940	75,000	7.00
1941	60,000	8.75
1942	18,000	15.75
1943	6,800	19.25

tecting alligators would harm the commercial alligator industry. (Not protecting alligators would benefit the industry—until there were none to kill.) One report noted that only two states had adequate laws to safeguard alligators in 1960. In that year the state of Louisiana prohibited the taking of alligators less than five feet long, and for the first time gave the Louisiana Wildlife and Fisheries Commission the authority to professionally manage the species. It was not enough, even with newly instituted short seasons. Poachers who killed alligators outside the legal framework were able to dump their skins on the market during the legitimate season, and populations failed to recover. The Bayou State banned the killing of all wild alligators in 1962 and two years later began a comprehensive research program on the ecology, reproductive biology, and captive propagation of alligators. Pioneer researchers like Robert Chabreck, Ted Joanen, and Larry McNease led the effort to decipher the reptile's life history. In Florida, the state with the second largest population of alligators, the situation was similar, and they also closed the season in 1962. Soon, alligators had legal protection throughout their range as other states enacted total prohibition of harvests.

The immediate result of barring the sale of alligators was a dramatic increase in the value of skins. Many hunters and buyers continued to operate because the penalties were negligible when weighed against the potential profits. Often fines could be recouped in one night of alligator hunting. Rampant poaching persisted through the 1960s until the amendment of an old federal law threatened violators with more than a slap on the wrist.

In 1900 the U.S. Congress passed what has become one of America's most potent weapons in the fight to prohibit illegal trafficking of wildlife. The Lacey Act was the first significant federal wildlife law in the country. At the time the few wildlife laws that existed were state regulations that could only be enforced within the enacting state's boundaries. If, for instance, a poacher illegally killed a thousand ducks in Maryland and sold them to a market in Virginia, he could not be prosecuted in Maryland after leaving the state. The Lacey Act closed the loophole by making it a violation of Federal law to cross state lines with illegally taken wildlife. The initial act only applied to birds and mammals. Some fish were added in 1926, but the amendment that affected alligators occurred in 1969 when reptiles, amphibians, mollusks, and crustaceans were tossed under the blanket of protection. At the same time penalties for misdemeanor violations of the law were increased to a $10,000 fine and/or one year imprisonment, and the maximum felony penalty became a $20,000 fine and/or imprisonment of five years. The law effectively prohibited all interstate movement of alligator skins. Since it was impossible to operate a poaching ring (comprising hunters, wholesale buyers, product manufacturers, retail sellers) completely within the boundaries of one state, the new Lacey Act became a serious threat to violators.

Alligators were soon to benefit from an even more powerful legal tool. Three years before the favorable Lacey Act amendments, Congress passed the Endangered Species Preservation Act of 1966. The Secretary of Interior was ordered to compile a list of endangered fish and wildlife and to preserve the habitats of those species on federal lands. An ominous honor, the list included the American alligator. It soon became obvious that this law did not go far enough to provide relief to imperiled species, however, and under the Richard M. Nixon administration a completely rewritten law was born—the Endangered Species Act of 1973. Those animals on the earlier endangered species list immediately came under federal protection wherever they were found. For alligators the tide began to turn.

The U.S. Fish and Wildlife Service (USFWS) together with state wildlife agencies implemented plans to restore alligators across the Southeast. Law enforcement efforts to protect the species intensified. Alligators from healthy populations were relocated to areas where they were scarce or no longer found. In 1979, one such project involved capturing alligators of all sizes on Sabine NWR in southwestern Louisiana and moving them to previously occupied range in northern Louisiana, Arkansas, Mississippi, and Alabama. During that experiment I hauled alligators to Arkansas's Holla Bend NWR on the northwestern edge of their historical range. State agencies monitored the convalescing populations and the USFWS began to relax stringent regulations where appropriate. Tightly controlled hunts with strict tagging regulations that followed the harvested alligator from the marsh to the manufacturer first began in Louisiana. In Florida the growing alligator population instigated increasing numbers of complaints about problem animals. In response, USFWS permitted a harvest of nuisance alligators. At varying rates, often depending on habitat quality, alligator populations continued to grow, and in 1987 USFWS pronounced the American alligator fully recovered and removed the animal from the list of endangered species. The revival of the alligator as a key component in wetland ecosystems of the Southeast is one of America's premier conservation success stories.

Modern Management

Alligators are currently classified as "threatened due to similarity of appearance" to other imperiled crocodilians. This status allows states to manage alligator populations with some federal oversight to assure protection of species such as the American crocodile that might otherwise get illegally caught up in commercial trade. Alligator management generally falls into one of three categories: nuisance alligator control, sustainable public hunting and trapping, and commercial alligator farming/ranching. Depending on the size of the alligator population and other factors, some states implement all three programs while others are less involved.

The dramatic recovery of alligator populations coincidentally occurred about the same time that many people began moving to coastal areas seeking aesthetic pleasures and a favorable climate. As the numbers of humans living in alligator habitat increased, so did the number of conflicts. The scale of the incidents is reflected in the total of nuisance alligator complaints received by the Florida Fish and Wildlife Conservation Commission (FWC) in 2010: 14,418. Louisiana, with more alligators and fewer people than Florida, still receives over 2,200 complaints annually. American history reveals that the threat of large predators, imagined or real, is rarely tolerated, and state natural resource agencies responded to complaints by establishing nuisance alligator control programs.

A nuisance alligator is generally defined as one at least four feet in length and posing a threat to people, pets, or other property. Nuisance control programs in Florida, Louisiana, Texas, Georgia, and South Carolina allow specially permitted hunters to remove problem alligators. State wildlife officials in other states handle the nuisance animals. In most cases, problem alligators are killed instead of relocated because of safety concerns. Regarding the 14,418 Florida complaints mentioned above, over 5,800 alligators were killed. The number of animals killed in nuisance control programs is considered when setting quotas for hunting/trapping programs. Several states have alligator nuisance control program objectives similar to those of Louisiana that include:

1. To minimize/alleviate alligator/human conflicts
2. To manage a statewide network of nuisance alligator hunters
3. To receive and process nuisance alligator complaints
4. To assign complaints to nuisance hunters
5. To ensure hunter compliance with nuisance alligator policy
6. To review and analyze nuisance alligator complaints and harvest data annually

The United States is party to the Convention on International Trade in Endangered Species of Wild Fauna and Flora (CITES). This agreement between many countries aims to ensure that the survival of wild plants and animals is not threatened by international trade. Federal regulations emanating from CITES require each state that conducts an alligator hunting/trapping program where skins are sold on the international market to demonstrate that such action is not detrimental to the overall alligator population. Since almost all skins are processed overseas, all states are impacted. States that allow alligator hunting/trapping (all except Oklahoma and North Carolina) meet the CITES obligations by routinely monitoring the status of their alligator populations and filing annual reports.

Monitoring involves counting alligators, a technique that is four-fifths science and one-fifth art. Surveys consist of two types depending on the habitat. In coastal marshes, aerial nest counts by airplane or helicopter are conducted along established transects in June. In wooded swamps, lakes, rivers, and bayous, nocturnal counts of individual alligators are made from boats using spotlights, again on established routes to ensure uniformity of data from one year to the next. The "art" aspect of counting involves the wildlife biologist's practiced skill at locating nests and alligators, and estimating the size of observed animals. Most states divide their alligator habitat into management zones based on habitat type and/or size of the population. During the surveys, habitat conditions such as water levels and vegetation are documented. When these data are compiled and compared with previous years' statistics, population trends can be noted. Harvest quotas are then determined for the upcoming alligator season and the appropriate number of tags issued accordingly. In order to meet CITES obligations each state's annual hunt plan must show that the proposed harvest will not be detrimental to alligators from a population perspective.

Regulations regarding alligator hunting/trapping programs vary from state to state. Land ownership often determines how alligator harvest tags and permits are issued. A private property owner (or his representative) may be issued tags if biologists verify that the alligator population on the tract can sustain a controlled harvest. Hundreds of thousands of acres of alligator habitat lie within the boundaries of public property. In this case, harvest tags are often dispensed via public lotteries. Individual states also define the seasons, legal methods of harvest, hunting hours, and tagging requirements. Seasons tend to fall in the months of August to October, well after the nesting period. Harvest methods include the use of firearms, archery equipment, set hooks, and snares among others. Some states allow alligators to be killed only during the day, others only at night, and others do not restrict hunting hours during the season. All require a CITES tagging system. A brief summary of recent alligator hunting laws by state follows, with the source being each state's published 2012 regulations. The complete regulations are comprehensive, detailed, and change frequently as wildlife agencies adapt policy to dynamic alligator populations.

Louisiana

General. Harvest tags are issued to owners of private lands and via lottery or auction to hunters for designated public lands. Harvest is permitted statewide in suitable habitat.

Seasons. State is divided into east and west hunting zones. The east zone opens the last Wednesday of August and the west zone opens the first Wednesday of September. Each zone remains open for 30 days from the opening date.

Hunting Hours. Alligators may be harvested between official sunrise and sunset only.

Limit. The daily and season quota is equal to the number of alligator tags that a licensed alligator hunter possesses.

Size Restriction. There are no size restrictions.

Legal Methods. Alligators may be harvested by hook and line, bow and ar-

Louisiana hunters with alligator caught on set hook. Photo by Paul Provence, U.S. Fish and Wildlife Service.

row, and firearms (except shotguns). The fishing (hook and line) method is the most common.

Florida

General. Each year, alligator management units are established with appropriate harvest quotas. Tags and permits are issued via a public lottery.

Seasons. Harvest periods are from September 12 until November 1, and from either August 15 until August 22, or August 22 until August 29, or August 29 until September 5, or September 5 until September 12, except as otherwise provided in the harvest permit.

Hunting Hours. Hunting is permitted from 5:00 p.m. until 10:00 a.m. during the season.

Limit. The limit is two alligators per permit.

Size Restriction. Only nonhatchling alligators may be taken.

Legal Methods. Alligators may be taken only by the use of artificial lures or baited, wooden pegs less than two inches in length attached to hand-held restraining lines or restraining lines attached to a vessel and hand-held snares, harpoons, gigs, snatch hooks, and manually operated spears, spearguns, crossbows and bows with projectiles attached to restraining lines.

Texas

General. Alligator hunting is permitted in two defined regions. In the first, 22 counties are designated "core" counties. In these counties tags are issued to landowners. All other counties where alligator hunting is permitted are considered "non-core" counties.

Seasons. The season in core counties is September 10–30. The season in non-core counties is April 1 through June 30.

Hunting Hours. Lawful hunting hours are from one-half hour before sunrise to sunset.

Limit. The limit for core counties is one alligator per unused tag. The limit for non-core counties is one alligator per person per season.

Size Restriction. There is no size restriction on alligators taken.

Legal Methods. Alligators may be taken with the aid of gigs, archery equipment, snares, hook and line, and firearms. Each method is specifically defined in the regulations.

South Carolina

General. The state is divided into four management units with 300 tags issued annually per unit by public lottery.

Seasons. The season runs from noon on the second Saturday in September through noon on the second Saturday in October.

Hunting Hours. Hunting is permitted at all hours during the season.

Limit. The limit is one alligator per permitted hunter.

Size Restriction. Only alligators four feet in length or longer may be taken.

Legal Methods. It is unlawful to kill or attempt to kill an unrestrained alligator. Alligators must first be captured alive prior to shooting or otherwise dispatching the animal. In order to capture an alligator, the hunter must first secure a restraining line to the animal. This may be accomplished by using handheld snares, harpoons, gigs, arrows, or snatch hooks. The alligator may then be killed using a handgun or bangstick.

Georgia

General. The state is divided into nine alligator management zones.

Seasons. The season runs September 1 through October 7 [2012].

Hunting Hours. Hunting is permitted at all hours during the season.

Limit. The limit is one alligator per permitted hunter.

Size Restriction. Only alligators four feet in length or longer may be taken.

Legal Methods. Hunters may use hand-held ropes or snares, snatch hooks, harpoons, gigs, or arrows with a restraining line attached. Legal alligators must be dispatched immediately upon capture by using a handgun or bangstick, or by severing the spinal cord with a sharp implement.

Mississippi

General. The state has six alligator hunting zones with a total of 810 permits issued via public lottery.
Seasons. The season runs from noon on September 7, until noon on September 17.
Hunting Hours. Hunting is permitted at all hours during the season.
Limit. Each person receiving an Alligator Possession Permit is allowed to harvest two alligators.
Size Restriction. Only alligators four feet in length or longer may be taken, and only one of the two alligator limit may exceed seven feet in length.
Legal Methods. Alligators must be captured alive prior to shooting or otherwise dispatching the animal. It is unlawful to kill an unrestrained alligator. Capture methods are restricted to hand-held snares, snatch hooks, harpoons, and bowfishing equipment. All alligators must be dispatched or released immediately after capture. Alligators captured with a harpoon or bowfishing equipment may not be released.

Alabama

General. The state has three designated alligator hunting areas with a total of 295 tags issued annually via public lottery.
Seasons. The seasons for Mobile/Baldwin Counties and west central Alabama are August 16–18 and August 23–25; the season for southeast Alabama is August 10–26.
Hunting Hours. The hunting hours are 8:00 p.m. (CDT) until 6:00 a.m. (CDT)
Limit. The limit is one alligator per permitted hunter.
Size Restriction. Only alligators six feet in length or longer may be taken.
Legal Methods. It is unlawful to shoot at or kill an unrestrained alligator. Restrained is defined as an alligator that has a noose or snare secured around the neck or leg in a manner that the alligator is controlled. Capture methods are restricted to hand-held snares, snatch hooks, harpoons, and bowfishing equipment. All alligators, if legal, must be dispatched immediately after capture or released. Firearms used for dispatching an alligator are restricted to shotguns with shot size no larger than no. 4 and bangsticks chambered in .38 caliber or larger.

Arkansas

General. The state has five alligator management zones. Hunting is permitted in two zones. A total of 36 permits are issued via public lottery [2011].

Seasons. The seasons are September 16–19, and September 23–26.

Hunting Hours. Alligator hunting is allowed from 30 minutes after sunset until 30 minutes before sunrise during the approved alligator seasons.

Limit. The limit is one alligator per permitted hunter.

Size Restriction. Only alligators four feet in length or longer may be taken.

Legal Methods. Alligators must be snared or harpooned and subdued using a hand-held snare or harpoon and hand-held restraining line before dispatching. The use of any other equipment is prohibited. Once subdued, an alligator may be dispatched using only a shotgun or shotgun shell-loaded bangstick using shot no larger than no. 4 common shot.

. . .

The following figures denote a recent annual harvest of wild alligators (excluding those killed in nuisance control programs) by state:

Louisiana (2011):	32,203
Florida (2010):	9,405
Texas (2011):	1,502
South Carolina (2011):	472
Georgia (2011):	219
Mississippi (2011):	182
Alabama (2011):	158
Arkansas (2011):	23

Alligator farming as a commercial enterprise began as early as 1891 in Florida. An 1894 publication described the situation then: "Alligator hunting as a profession has almost died out in Florida. The farms where 'gators are reared, cared for, and when matured, killed, with the tanning pits and house close at hand, have pushed the one-time famous Alligator Cracker to the wall, and he is gradually becoming extinct as the Dodo from mere drifting into other lines of sport." A 1905 account describes the Indian River operation of a locally well-known character, Indian Joe:

There are several sources from which the promoter of the alligator farms expects to derive the revenue to recompense him for his outlay. The chief of these is the sale of live specimens. There is a constant demand for small 'gators from private collectors, and for the full-grown animals from city aquariums and menageries. Another form in which they are salable is as stuffed specimens, and a third demand is for alligator hide to be manufactured into leather. This last has not yet reached the point when it is profitable to raise 'gators for their hides alone, for the Seminole Indians of the Everglades, who exist almost entirely by the hunting of alligators, are able to sell the hides at two or three dollars apiece.

A different writer describes a farm on the edge of the historical range of the species:

About seven years ago [1903], H. I. Campbell, son of a British Colonel, and a well-known alligator-hunter, established at Hot Springs, Ark., a farm for the propagation of alligators for commercial purposes. The strange farm proved successful from the first, and was extended until it covers several acres, on which there are from five hundred to eight hundred alligators constantly. The Arkansas farm is situated on the banks of a small stream, which forms several ponds and lakelets that serve as excellent breeding-places for the "stock." Alligators of all ages and sizes may be seen there, from babies as small as lizards to aged monsters.

This same source tells of another remarkable farm 1,200 miles from the nearest natural alligator habitat. Started in 1907, a venture near Eastlake Park in Los Angeles had breeding animals along with incubation and hatchery facilities. It was not until almost 75 years later, however, that a viable industry developed around the husbandry of American alligators.

Modern commerce began on a small experimental basis as techniques for efficiently rearing large numbers of alligators in a controlled environment were worked out. Two basic approaches evolved. The first is to have a completely self-contained system in which a producer maintains his own breeding stock, hatchery, and grow-out facility. Eggs are harvested from nests of captive alligators, artificially incubated, and the young alligators are then raised to a marketable size. The second technique eliminates the necessity of keeping breeding stock. Eggs or young alligators are collected from the wild or purchased from specialty farms or state agencies. (This method is often called "ranching," but more commonly all approaches are now termed "farming.")

Most alligator farming today is based on the collection of eggs from wild alligator nests because of the overall poor hatching rate of eggs laid by captive animals. The reason for the differences are unclear but may be tied to the role of fatty acids in embryo development. Researchers in one study analyzed the contents and then compared the hatching rate of fertile eggs from wild and captive alligators. A striking 94 percent of the wild eggs hatched as compared to 50 percent of the captive eggs. The yolks of the eggs had contrasting ratios of certain types of fatty acids critical in embryo development, perhaps because of differences in their diets.

The concept of using wild alligator eggs to support commercial activities was initially and justifiably met with skepticism by some biologists, conservation organizations, and members of the public. Before the practice became commonplace, questions would need to be answered. The Florida Game and Fish Commission listed their prerequisites as follows:

· "The long-term impact of such a harvest on alligator populations would need to be examined."
· "The resource had to be equitably distributed among potential users."
· "It would be essential that ranching provide some positive economic feedback to the wetlands which supported the wild populations."

Of paramount importance was the necessity that the egg-collecting programs not have a negative impact on wild alligator populations. The general idea was to collect the eggs, have the farmers incubate and hatch them, raise the young animals to a predetermined size, and then release a portion of them back to the areas where the eggs were gathered. How many eggs should be collected from the wild? What percentage of the hatched alligators should be returned to the wild? How big should the young alligators be when they are released? Can the released alligators survive in the wild after life on the farm? Beginning in the 1970s, state biologists, mostly in Louisiana and Florida, began intensive research programs to answer these questions and more.

One important four-year study of juvenile wild and captive-raised alligators that had been released to the wild took place at Rockefeller Wildlife Refuge in southwestern Louisiana. The results showed that the farm-released alligators grew at least as well as the wild ones and many grew significantly better. Other observations included that males in both groups grew faster than females; the rate of growth in wild alligators slowed as they grew larger, but growth rates of farm-released alligators remained accelerated even at the larger size classes; the projected maximum length of farm-released alligators was greater than wild ones; the general body condition of animals in both groups were the same.

Alligator farming today is a good example of adaptive management of natural resources. This process involves trying different management practices, evaluating the outcome, and adjusting the program until desired objectives are obtained. In its simplest form, it is planning ahead to learn from mistakes. To protect the resource, the approaches begin conservative and are fine-tuned as data are accumulated. For example, the Louisiana Department of Wildlife and Fisheries initially required alligator farmers to return 17 percent of hatchlings to the wild. After several years of research proved that released alligators are able to feed in the wild, maintain high rates of survival, and successfully reproduce, the restock rate was reduced to 14 percent. At 12 percent the rate now sits at a level documented by research to be sustainable without reducing the wild alligator population. (This is based on the belief that only 12 percent of alligators that hatch in the wild survive to reach 3 to 5 feet in length—the length at which farmed alligators are returned to the wild.) Other investigations have determined the best methods to collect eggs and the optimum size of released alligators. Biologists decide the number of eggs that can be collected annually by routine nest and alligator surveys.

On a cool April morning I visited a modern Louisiana alligator farm that is typical of many others. (Variations in businesses are a result of farming objectives, site location, and size of facilities.) Established in 1987 this farm has evolved over the years into a smooth-running operation. The owner narrated a recent average year's work and added that his figures are approximate.

The process begins when he engages with landowners to collect eggs on their property. Contracts are negotiated on an annual basis, but he usually gathers eggs on the same properties every year. He collects 25,000–30,000 eggs annually on 150,000 acres. About 130,000 of these acres are coastal marshlands with the remainder scattered across several interior wetland tracts comprising swamps and lakes. The landowner is paid an average of $5.00 per egg.

Using a helicopter, he locates nests in the marsh and marks them with GPS coordinates to facilitate efficient collecting. Timing of egg gathering is important. Egg mortality is unacceptably high (35 percent) if collected in a critical period when they are 7 to 17 days old. The goal is to collect the eggs before or after this precarious stage of development when the hatch rate is about 85 percent. Eggs are candled to determine their age, and if the time is right, the entire clutch is gathered. Because alligator embryos attach to the tops of eggs, embryos will die if eggs are turned over. For this reason, the tops of eggs are marked so they can be kept in the proper position. The clutches are placed in individual plastic milk crates along with vegetation from the nest. The location, age, and other data pertinent to the nest are recorded. Most eggs are collected with the aid of airboats, and a 5- or 6-man crew spends 20 to 30 days per year collecting eggs for the farm.

The crates of eggs are trucked to the farm and placed on racks in a high-ceilinged hatchery building that functions as an incubator. Heaters and a misting system regulate the temperature and humidity. The farmer would prefer a majority of male alligators because they grow somewhat faster, so the incubator is set at 91°F. After about 65 days from laying, the eggs begin to hatch. Baby alligators emit their high-pitched croaking when hatching, which alerts technicians who are closely monitoring the incubator. When some eggs in a clutch begin to hatch, technicians open all of the eggs in that group. The babies remain attached to their amniotic sacks for one day to allow for nutrient absorption and are afterward often treated with an antibiotic such as tetracycline to prevent bacterial infections.

Young alligators are moved from the hatchery to a grow-out building. Such facilities come in many different sizes and designs, but all must meet the specifications of various regulatory agencies. All are insulated and heated and contain water filtering, circulation, and drainage features. At the farm I visited the grow-out buildings resembled midsize, commercial chicken houses that had been partitioned into stalls. Each stall was completely walled off and contained a fiberglass-lined concrete vat capable of holding water to a depth of two feet. Alligators are fed on wooden trays suspended just above the water. A flushing system sends all waste to a nearby system of treatment ponds. The number of alligators per stall varies according to the size of the animals.

Hatchlings at the farm are first fed high-protein, pelletized food and kept at 89–90°F. The owner reports that most developmental and disease problems occur in the first month after hatching. At three months of age the alligators receive food with slightly less protein content, and at eight months the stall temperature is lowered to 83–84°F to reduce aggressiveness. (Even relatively small alligators can be injured in fights with their cohorts. Scarring also greatly reduces the value of a skin.) By this time they are fed a ground mix of commercial pellets and chicken parts.

The farm owner states that many of his colleagues harvest their alligators at 12 to 16 months of age when they are 3½ to 4 feet long. Referred to as "watch-strap" size, most such skins are made into watch bands. He, on the other hand, raises 90 percent of his stock for the handbag and belt market, which requires 6-foot skins and 3 years of growth. Since alligators of various sizes are always on the farm, some harvesting goes on year round.

The alligators are processed in a large, clean, on-site facility not unlike a typical meat-packing plant. They are killed, cooled down in water-filled vats, and then iced for skinning the next day. After the labor-intensive skinning procedure on stainless steel tables, the hides are placed back in water until they are pressure-washed to remove all flesh from the inside of the skin. The hides are then packed in salt and rolled up ready for sale.

The farm produces 15,000–20,000 skins annually. Most are purchased by a French company that has a tannery in Lafayette, Louisiana. When skins are delivered to the plant, they are measured, graded, and valued. Some are tanned there, and others are shipped raw overseas for tanning and manufacturing into finished products.

At the time of my visit, alligator meat was in high demand with the wholesale price at $7–$8 per pound. The farm sells meat to a buyer who prefers whole skinned and gutted carcasses ready to be processed into various products such as steaks and sausage. The harvested alligators yield 30–40 pounds of meat each resulting in an annual crop of 300,000–400,000 pounds.

In addition to farmed alligators, the owner buys several hundred wild alligators from hunters during the annual state hunting season. He also purchases nuisance alligators throughout the year from permitted nuisance alligator trappers.

From a conservation perspective the most important event in the annual cycle of an alligator farm is the obligated release of alligators back into the wild. For the farmer, too, a successful release that fulfills his permit requirements is critical to his operation. In Louisiana an official from the Louisiana Department of Wildlife and Fisheries visits each farm to inspect the alligators that are proposed for release. They are measured, sexed, tagged with web tags, and tail-notched by year class. Each year the farm that I visited returns 2,000–2,500 alligators 3 to 5 feet long back to the areas where they were collected as eggs.

Alligator farming has evolved as farmers and researchers from universities and conservation agencies have developed improved techniques to enhance the efficiency of alligator production. Pelletized feeds are cleaner and eliminate costs associated with maintaining frozen meats. Advanced designs of grow-out buildings using stacked pens permit growing more animals in a smaller space, thus reducing energy costs. Water recycling often occurs. Advances in veterinary medicine as it applies to alligator farming are ongoing. Still, anxieties exist. When I asked the owner about his greatest concerns for the alligator farming industry, he mentioned issues common to any business including market fluctuations such as the recent global economic recession. As luxury items, alligator hide products are especially sensitive to market volatility. Competition from foreign countries are a worry (the expanding crocodile farming industry in southern Africa is seen as a threat), as are other factors beyond his control (such as the recent season when almost all alligator nests on the Louisiana coast were destroyed by aberrant flooding, resulting in no egg collecting for a year). Diseases still threaten as evidenced by West Nile virus first diagnosed on a Florida alligator farm in 2002, and an antibiotic-resistant, mycoplasma bacteria that killed large breeding males on another farm.

The overall trend in alligator farming in recent years has been a decline in the total number of farms but an increase in the number of skins produced (with the exception of a downturn in 2010–11 that occurred because of low stocking rates in 2008). A summary of alligator farming information by state follows:

Louisiana. Louisiana dominates the alligator farming industry. Beginning in 1972 when 35 farmed skins were sold, production increased until a record high of 305,176 skins were marketed in 2007. The value of the 2011 farmed alligator crop was estimated at $38.5 million. Currently there are 32 licensed alligator farmers active in the state, and 60 licensed dealers buy/sell skins, meat, heads, feet, and teeth (some or all). Louisiana farmers obtain over 90 percent of their eggs from the wild. In 2011, 353,176 eggs were collected of which 300,546 hatched. Farmers returned a total of 14,357 subadult alligators to the wild. (Note: farmers have 24 months from the year of egg collection to make required returns.)

Florida. Florida's alligator farming industry ranks second and is considerably larger than lower ranked states. Approximately 28,000 farmed skins are produced annually. The net sales of alligator hides and meat fell from a high of nearly $7 million in 2008 to $2.6 million in 2010. The average value of a marketed farm alligator in that year was computed at $140.98. Licensed farms in 2010 totaled 59, but only 15 were active. Florida farmers procure about 75 percent of their stock from wild eggs and hatchlings. The other 25 percent are raised from captive breeders. About one-third of Florida farms have at least some breeding stock. Many eggs and hatchlings are sold to farms in Louisiana and Georgia.

Other states. The extent of alligator farming in other states compared to Louisiana and Florida is small. Texas has 12 licensed alligator farms, Georgia has 10, and Mississippi has 3. Most get their eggs or hatchlings from Louisiana or Florida. South Carolina legalized alligator farming for several years, but no permits for the activity were requested and the statute expired due to lack of interest.

. . .

The sustainability of farming alligators using eggs from wild nests is now proven, at least in the relative short-term. Information gathered from tagged, farm-released alligators shows that they feed, grow, survive, and most important, reproduce successfully in the wild. Research in Florida and Louisiana demonstrates that most of those animals harvested by farmers (via eggs) would not have survived in the wild due to natural mortality. Moreover, wild alligator populations in those states with egg removal programs are stable or increasing slightly. Perhaps the most important outcome of this industry is the incentive it gives private landowners to maintain and protect good alligator habitat. Many species of plants and animals benefit from the economic value of alligators through the sound stewardship of wetlands.

8

A Love/Hate Relationship

A Cultured Creature

Alligators have permeated our society and culture in ways mundane and exotic, practical and bizarre. Geographic names alone give some indication of our fascination with this species. The term "alligator" and its synonyms embellish maps across the country. The *Omni Gazetteer of the United States of America* lists almost 200 place-names containing the word alligator. These include 108 rivers, streams, and bayous, 59 lakes, and 20 swamps. Alligator, Mississippi, is a village of about 200 people in the northwestern county of Bolivar. Alligator Alley is a highway that traverses the Everglades across southern Florida. And there are 27 states that have an alligator place-name with Florida leading the way with 63. Texas has 48, Louisiana has 40 (although Louisiana also has many place-names containing the Cajun French synonym "cocodrie"), and North Carolina has 30. Even some states far outside the historical range of the species have alligator place-names. Surfers seek a persistent wave in Oahu, Hawaii, near Alligator Rock. Similar labels show up on the maps of Alaska, Idaho, Nevada, New Mexico, Oregon, Wisconsin, and Wyoming.

Within the commercial realm, the diversity of businesses that use the term alligator in their names, logos, and advertising seems unlimited. Alligator Records produces blues music, Alligator Cable makes brake lines, and Alligator Brewing turns out zesty spirits. In the fashion industry, Alligator Sunglasses deflect swamp glare and the crocodile logos on Izod/Lacoste shirts became alligators on the American side of the Atlantic Ocean. At the turn of the twen-

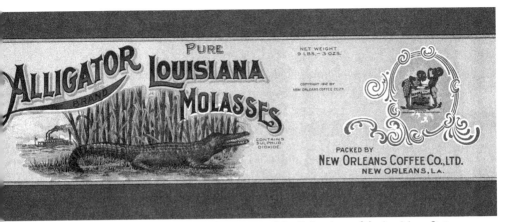

Early twentieth-century molasses label. Courtesy of the Collections of the Louisiana State Museum.

tieth century the Alligator Hotel in Dahlgren, Illinois, was the place to stay and included a pool of live, annually replenished alligators in the courtyard. Today the Mermaid and the Alligator Bed and Breakfast in Key West is popular with tourists. Across the landscape, dining and libations are available at an array of alligator restaurants, grills, cafes, and lounges. "Gator," the truncated version of the reptile's name is no less popular. Gatorboats jet across wetlands, and many collegiate and professional athletic teams consider themselves at a disadvantage if the sport drink Gatorade is not on their sideline.

Anthropomorphic versions of alligators are well-represented in mascots. None are more famous than Albert E. Gator and Alberta Gator of the University of Florida in Gainesville. Originally characterized by live alligators, Albert and Alberta exist as plush costumes manned by heat-tolerant students today. Many high schools have also adapted alligators as mascots. In a manner, the vogue has even gone extraterrestrial. When the seven astronauts of space shuttle *Endeavour*'s STS-127 flight in 2009 were delayed from reaching the launchpad by a large alligator crossing the road, the crew adopted the creature as their mission mascot.

Our music has not escaped the reach of alligators. One fan compiled the following list of popular songs and artists:

Top 10 Alligator Songs

1. "See You Later Alligator"—Bill Haley & His Comets
2. "Cold Night for Alligators"—Roky Erickson
3. "Alligator"—The Grateful Dead
4. "Alligator"—Grizzly Bear

5. "Alligator Pie"—Dave Matthews Band
6. "Alligator Chomp (The Ballad of Dr. Martin Luther Frog Jr.)"—Shooter Jennings
7. "Alligator Boogaloo"—Clarence "Gatemouth" Brown
8. "Alligator Crawl"—Fats Waller
9. "Alligator"—Tegan and Sara
10. "Alligator"—Leslie West

As characters in books, alligators are variously cast as protagonists, antagonists, and minor passersby that enhance a setting. Two of Joel Chandler Harris's Uncle Remus stories starred alligators. Two dozen modern children's books feature alligators in tales with moral and educational overtones. In novels, alligators are almost always depicted as cunning maneaters and dressed with just enough factual natural history to nurture the otherwise unrealistic image. With movies, the story is the same. From *The Alligator People* (1959) to *Gator Bait* (1973) to the James Bond film *Live and Let Die* (1973), alligators got a bad, inaccurate rap. Perhaps the most literary alligator was Walt Kelly's *Pogo* of comic strip fame. In a gumbo of wit and comedy *Pogo* dished out social satire from the Okefenokee Swamp from 1941 until the 1990s.

For years we have been enthralled with alligators in zoos, menageries, and

Alligator skin products. Photo by Louisiana Alligator Advisory Council.

sideshows. We photographed them there, at one time "wrestling" with their keepers or in other ridiculous poses, and plastered the images on postcards for folks back home who were not so lucky to see such marvels. We love them no less in souvenir shops and high-end boutiques, purchasing jewelry made from their teeth and curios crafted from their stuffed babies. With their tanned hides we cover furniture and books, and fashion the skins into trunks, gun cases, saddles, wallets, music rolls, and $12,000 Louis Vuitton handbags (retail price of vintage Saint Cloud model). Along the way, we consume them in recipes entitled "alligator etouffee," "sweet and sour alligator," and "alligator primavera."

To humans, alligators are so unique in appearance and behavior that we use their moniker to name other living animals and plants—and inanimate objects. Consider in the animal kingdom the alligator gar, alligator lizard, and alligator snapping turtle. As for the botanical arena, representatives include alligator pears, alligatorweed, and alligator juniper. In the man-made matter department, there are alligator clamps and alligator wrenches among others.

The most definitive modern example of human fascination with the American alligator is exposed in the viewing statistics of the reality television series entitled *Swamp People*. First broadcast on the History Channel on August 22, 2010, the program tracks the daily activities of several rural alligator hunters through the swamps and bayous of south Louisiana. The series premiere reaped 3.1 million viewers, an average that held up throughout the season. In 2011 the average viewers increased to 4.1 million, and the final episode broke viewing records and ratings by attracting 5.5 million viewers. Billboards advertising the program adorned Times Square in New York City. Wildlife managers across the Southeast have told me that, as a spinoff of the program's success, they are experiencing an increasing demand for permits to hunt alligators. The attraction lies not in the natural history of alligators, but rather in the highly drama-tized human interactions with the species. Ironically, the realism viewers are exposed to as it pertains to harvesting alligators is limited to a basic framework of the process. Nevertheless, no single species of wild animal has ever received such significant airtime, and even if authenticity of the show is less than com-plete, the conservation ledger is likely enhanced by the unwitting absorption of at least some ecological knowledge by several million people.

The Threat—Real and Imagined

Being consumed by a wild animal ranks as one of the most primal of human fears. In our vivid imaginations few horrors are more dreadful to contemplate. Fear is not an emotion exclusive to humans. In animals, too, fear is real and measurable (by changes in corticosteroid levels). Although fear is primal, it may not necessarily be innate, at least in higher life forms such as mammals. Recent research suggests that a fear of predators is not instinctive but is learned behavior that only occurs when prey species are exposed to animals that routinely eat them. Examples include Yellowstone elk that lived in a wolf-free environment for many years until the late 1990s. When wolves were reintroduced to the ecosystem, elk showed little fear of them at first but soon modified their behavior to survive. If the premise is applicable to humans, then it begs the question of why are we afraid of alligators often at a level inconsistent with the actual threat.

In spite of exaggerated European accounts, there is no evidence that alligators were ever more than a minor environmental menace requiring reasonable caution by Native Americans. The Old World perspective at the time may have had origins in known crocodile behavior. Nile and saltwater crocodiles are thought to attack and kill more humans than any other predatory animal. For alligators, the verdict may have been "guilty by similarity of appearance." Pre-twentieth-century reports of alligator attacks are fairly common, and some no doubt have basis in fact. One of the first recorded instances involves a member of the last, ill-fated La Salle expedition who is said to have been devoured in southeast Texas in 1686 as he swam a bayou. The later experiences of a Dr. Kilpatrick in Concordia Parish, Louisiana, are typical:

> One of my neighbors, J. P. McCoy, was attacked by an alligator while fishing on a log over a lake, near here, in 1840. He was standing on a log which projected far out in the lake, and in an unguarded moment the monster sprang up and seized him by the right hand and arm, and by a rapid succession of gyratory wrenches, fractured the bone, and nearly separated the arm before he could be forced to let go. . . . McCoy's arm inflamed, and so endangered his life, that it had to be taken off above the elbow. . . . Another circumstance, more horrifying than this, occurred on Black River, in 18__. Mrs. _____ was washing clothes on the bank of the river, and had her child there with her, lying near the water, when suddenly an alligator sprung out, and seizing the child, swam with it to the other shore, where it leisurely proceeded to devour it. There being no boat at hand, the hideous monster finished his meal undisturbed, save by the unavailing screams and wailings of the agonized mother.

Even today in areas such as Florida with dense human populations superimposed on prime alligator habitat, predatory attacks by alligators on people

An indicator of human/alligator conflict. U.S. Fish and Wildlife Service, NCTC Image Library.

are uncommon. In the modern world the perpetuation of old myths is made easier by pervasive digital media and rampant urban myth.

So what are the actual numbers regarding alligator attacks on humans? Prior to the 1970s attacks were poorly documented and rarely the object of detailed investigations. As a result figures from different sources often vary. One report lists the following number of alligator attacks between 1948 and 2005 in each state within the species' range: Florida (337), Texas (15), Georgia (9), South Carolina (9), Alabama (5), Louisiana (2), Arkansas (1), North Carolina (1), Mississippi (0), Oklahoma (0). During this 57-year span, Florida attacks far exceeded all other states combined. (Florida also leads the United States in number of unprovoked shark attacks. The number of annual attacks between 2000 and 2010 averaged 23.4) An obvious anomaly lies in the fact that Louisiana with an alligator population nearly twice that of Florida had only two reported attacks. The contrast is simply a function of demographics. Florida's human population exceeds 19 million, and most people there live, work, and play in or near prime alligator habitat. Except along the Gulf Coast, Louisiana's 4½ million people are not as exposed to alligators. Since 1979 the Florida Fish and Wildlife Conservation Commission (FWC) has maintained

Table 6. Alligator Bites on Humans in Florida, 1979–2012

Year	Minor	Major	Fatal	Total
2012	5	2	0	7
2011	0	2	0	2
2010	0	3	0	3
2009	2	6	0	8
2008	1	5	0	6
2007	8	6	1	14
2006	4	8	3	12
2005	1	9	2	10
2004	3	9	2	12
2003	1	9	1	10
2002	5	9	0	14
2001	5	11	3	16
2000	5	8	0	13
1999	5	5	0	10
1998	2	3	0	5
1997	1	4	1	5
1996	2	4	0	6
1995	4	9	0	13
1994	6	5	0	11
1993	5	8	2	13
1992	1	6	0	7
1991	5	5	0	10
1990	10	3	0	13
1989	6	4	0	10
1988	3	6	1	9
1987	6	3	1	9
1986	10	3	0	13
1985	2	4	1	6
1984	0	5	1	5
1983	1	5	0	6
1982	3	4	0	7
1981	3	5	0	8
1980	0	4	0	4
1979	0	2	0	2

Note: All bites reported in this summary table are considered unprovoked bites, which are defined as bites on people by wild alligators that were not provoked by handling or intentional harassment. Minor bites are those in which the victims' injuries were superficial and required no treatment or only first aid. Major bites are those in which the victims' injuries required medical care, beyond first aid, to treat wounds. The "major" column includes fatal bites.

excellent records and investigated those cases that resulted in mortality. The following report from that agency is current as of June 2012:

Between 1948 and 2012, the FWC documented 344 attacks on humans in Florida. Of these attacks, 22 were fatal. At least 9 other cases involved individuals who may have been dead before the alligator attack. Published details of 22 recent cases that resulted in human mortality follow:

Fatal Alligator Attacks on People in Florida

1. Female, 16, killed while swimming in Oscar Scherer State Park (Sarasota County) on August 16, 1973, at dusk. The 11 foot 3 inch healthy male alligator had been fed by visitors.
2. Male, 52, was seized by the arm while swimming in the Peace River Canal (Charlotte County) on September 28, 1977, at 8:35 p.m. The seven-foot female alligator, which had appeared in the canal the day of the bite, severed the victim's arm at the elbow. The man died three days later of complications from the bite.
3. Male, 14, was killed while swimming across the Hidden River Canal off Bessie Creek (Martin County) on September 10, 1978, at noon. The alligator was an 11-foot healthy male.
4. Male, 11, was killed while swimming in a canal in St. Lucie County on August 6, 1984, at 4:30 p.m. The alligator was 12 feet 4 inches, aged and in poor health.
5. Male, 27, disappeared while diving and harassing small alligators in the Wellington C-27 Canal near West Palm Beach on May 4, 1985. His body was recovered two days later with severe injuries to the neck and puncture wounds on the arm.
6. Male, 29, was killed while snorkeling in the Wakulla River on July 13, 1987, at 2:00 p.m. The alligator was an 11-foot healthy male.
7. Female, 4, was seized and killed by an alligator while walking along the shore of Hidden Lake (Charlotte County) on June 4, 1988, at 6:10 p.m. The alligator was a 10 foot 7 inch male.
8. Male, 10, was killed while wading in the Loxahatchee River at Jonathan Dickenson State Park (Martin County) on June 19, 1993. The alligator was an 11 foot 4 inch male.
9. Female, 70, was killed at Lake Serenity (Sumter County) on October 3, 1993. The circumstances surrounding her death are unknown, but she died of a broken neck caused by an alligator bite to the throat and head.
10. Male, 3, was killed at Lake Ashby (Volusia County), on March 21, 1997. The child strayed outside the roped-off swimming area in a

county park to pick some lily pads when an 11-foot alligator attacked him. Splashing dogs in the area may have attracted the alligator.

11. Male, 70, was killed in a pond near his residence in Venice (Sarasota County). He was found on May 4, 2001, and the county medical examiner determined that he died from multiple trauma and loss of blood. An 8 foot 4 inch alligator was destroyed.

12. Female, 2, was killed at Lake Cannon (Polk County) on June 23, 2001. She wandered 700 feet from her fenced backyard where she had been playing when last seen by her mother. A 6 foot 6 inch alligator was destroyed.

13. Male, 82, was killed in Sanibel on September 11, 2001. He was walking his terrier on a narrow path that ran between two wetland areas when a 10 foot 9 inch alligator seized him and dragged him into the water, severing his leg. FWC Officers destroyed the alligator.

14. Male, 12, was killed while swimming near a boat ramp in the Dead River (Lake County), on June 18, 2003. The male alligator that attacked and drowned him was 10 feet 4 inches long and weighed 339 pounds. That alligator and several other large alligators were captured and destroyed.

15. Female, 54, was seized by an alligator while landscaping near a pond along Poinciana Circle, Sanibel, on July 21, 2004. She survived the attack but died later of an infection related to the wounds. The alligator that attacked her, a 12 foot 3 inch male, was destroyed.

16. Female, 20, was killed while swimming after midnight in a retention pond at the Lee Memorial Health Park (Lee County) on September 26, 2004. The 7 foot 11 inch male alligator that attacked her was destroyed.

17. Male, 56, was found dead in Six Pound Pond near Lakeland (Polk County), with multiple alligator bites and the left arm amputated below the elbow. The medical examiner determined that the victim was bitten prior to dying on March 11, 2005. The 9 foot 8 inch male alligator responsible for the attack was destroyed.

18. Male, 41, was killed while swimming in a canal in Port Charlotte on July 15, 2005. The 12 foot 2 inch alligator that attacked him was destroyed.

19. Female, 28, was killed by an alligator at the North New River Canal in Sunrise on May 10, 2006. Circumstances of the attack are uncertain because no witnesses were present. The 9 foot 6 inch male alligator that attacked her was destroyed.

20. Female, 42, was killed by an alligator in a canal in the East Lake Wood-

lands subdivision in Oldsmar on May 13, 2006. Circumstances of the attack are uncertain because no witnesses were present. The 8 foot 5 inch female alligator responsible for the attack was destroyed.

21. Female, 23, was seized and drowned by an alligator in Juniper Run in the Ocala National Forest on May 14, 2006, while snorkeling. She had separated from others in her party and was alone when the attack occurred. Companions found her in the jaws of the alligator less than 30 minutes after the attack and forced the alligator to release her by assaulting its head. The 11 foot 5 inch male alligator was captured four days later and destroyed.

22. Male, 36, was seized and drowned by an alligator as he was swimming across a pond at the Miccosukee Indian Reservation in West Miami on November 8, 2007. Eyewitnesses watched as he disappeared under water while trying to elude police. Divers later recovered his body at the bottom of the pond. A 9 foot 4 inch alligator and a 7 foot 6 inch alligator were removed from the pond, the larger of which was believed to be responsible for the attack.

A cursory assessment of the tragedies reveal that most but not all victims were swimming or wading, many were young, most were unaware of the alligator's presence at the time of attack, and none were feeding alligators. The 22 attacks occurred during 8 different months of the year, and most of the predators were large males greater than 9 feet long. As the government agency with the most experience investigating alligator attacks, the FWC has released a fact sheet to inform and advise the public of safety measures. A summary of its contents follows:

General Information. Alligators seldom attack humans, and fatalities from such attacks are rare. However, confrontations between people and alligators are increasing as a result of the burgeoning human population and loss of wildlife habitat to development. Alligators are naturally afraid of humans, but they lose that fear when people feed them. Alligators that are fed learn to associate people with food. Biologists report that alligators seldom attack humans for any reason other than food. Although alligators are quite agile and swift on land for short distances, attacks are most likely to occur in water. Before an attack, an alligator may stalk its prey. Once the reptile decides to attack, it usually swims underwater to within a few feet of its victim and surfaces to clamp down on an arm, shoulder, or leg. Alligators prefer prey that they can overpower easily. The size of the alligator and the size of the prey are the primary factors that determine whether the reptile will attack.

Avoid Alligator Attacks. Alligators tend to feed mostly at dusk or during early evening. Avoid swimming in waters known to be inhabited by alligators during that time of day, particularly during summer months when alligators are most active. Since dogs and cats are similar in size to wild prey favored by alligators, they should not be allowed near the water in known alligator habitat. Pets on the shore may also draw alligators to areas used by swimming humans. Fish-cleaning waste can attract alligators. Swimming or even dangling feet in the water where fish have been cleaned can be very dangerous. Using your feet or hands to search water bodies in alligator habitat for objects such as golf balls is risky behavior.

Fight off an Attacking Alligator. If attacked the best tactic is aggressive resistance. Victims should fight the predator with all their being. A loud, forceful struggle often confuses the alligator and causes it to retreat.

It is little consolation to victims of alligator attacks or their families and friends that such events are rare when considered in the context of overall risks that we encounter in the routine affairs of life. Even when we limit threats to wildlife-related risks, the chances of dying from bee stings or in an auto accident caused by a deer collision are much higher than death in the jaws of an alligator. As mentioned earlier, the fear of alligator attack is also disproportionate to reality and may be self-inflicted as we reap the rumors of old tall tales and new internet blogs. For me the disparity of dread is also primal and tied to the fact that while deer and bees can be deadly, they cannot devour us in the manner of a large predatory alligator . . . and thus they thrive in our imaginations in the sewers of New York City, sometimes even above and beyond the real human-sourced dangers in that "ecosystem."

"What Good Is It?"

Medicine

Aldo Leopold is considered the founding father of wildlife conservation in America. A forward-thinking man with a deep understanding of human dependence on healthy, natural ecosystems, he once remarked that "the last word in ignorance is the man who says of an animal or plant: 'What good is

Using alligators as a tool for environmental education. U.S. Fish and Wildlife Service, NCTC Image Library.

it?'" Unfortunately, many people still ask that selfish question in regard to alligators. For those, answers laden with aesthetic values or the merits of "keeping all the parts" are at right angles with their reality. However, a professor who works in alligator country has an answer they can relate to.

Alligators have no love lost for their own kind. Territorial disputes and blatant acts of cannibalism often result in horrific open wounds. They live in a subtropical, aquatic environment teeming with a witch's brew of microorganisms ready to saturate the smallest scratch with potentially lethal pathogens. Yet they survive catastrophic injuries, even thrive afterward. This conundrum was not lost on a south Louisiana biochemist. Dr. Mark Merchant set out to determine why the alligator's immune system is so potent. He captured alligators, collected blood samples, and headed for the lab where the serum was separated and exposed to 16 different strains of bacteria. It killed them all, including *E. coli* and others that cause salmonella, dysentery, and staph and strep infections. As for viruses, it eradicated herpes simplex and one type of HIV. Even some kinds of disease-causing fungi are susceptible. Merchant said, "It takes down practically everything."

Scientists in the medical field are concerned about the increasing number of dangerous microbes that have developed resistance to current antibiotics. To stay ahead of mutating germs they are searching the world for new animal and plant sources of effective drugs. The exciting discovery of alligator blood properties was the precursor of research to determine the "how and why" of the blood's potency. Merchant and his colleagues began dissection of the serum by pulling out the white blood cells that attack disease organisms. They were seeking the chemical composition of proteins in the white blood cells. These short chains of amino acids, called peptides, do the dirty work, and once their chemistry is exposed may be reproduced artificially. At that point experimental antibacterial drugs can be produced. Already it is surmised that diabetics may use resulting creams to treat foot ulcers and avert amputations, while application to burn victims will prevent infections. In the form of pills the new medicine will fight internal infections. It is not a simple process but fruitful progress continues. Merchant's answer to those who question the practical value of alligators is, "There's a real possibility that you could be treated with an alligator blood product one day."

The Swamp Canaries

Alligators are apex predators that live at the top of the food web in their ecosystems. As such, they are particularly susceptible to environmental contaminants that increasingly accumulate in those organisms below their level in the web. The classic American example involves the bald eagle, another top predator, whose populations plummeted when the pesticide DDT caused their egg shells to be too thin to support incubation. The eagles acquired DDT from fish that in turn picked it up in their food source. Contaminants are detected in alligators at some level almost every time they are tested. DDT, DDE, DDD, dieldrin, heptachlor epoxide, lindane, and PCBs have been found in alligator tissues. Heavy metals, extremely toxic at certain levels, are also ubiquitous in alligators across their range. Arsenic, cadmium, lead, mercury, chromium, iron, zinc, and copper have been identified at abnormal concentrations in alligator blood. Mercury levels in alligator meat that is sold for human consumption has instigated a number of enlightening studies. High mercury levels have been noted in alligators from Georgia, Alabama, Florida, and South Carolina, often in amounts greater than is considered safe to eat for fish. Because of mercury concerns, alligator harvests and the sale of meat from nuisance alligators in parts of south Florida including the Everglades was prohibited for several years. These areas have since been reopened to recreational harvest, but the ban on meat sales remains.

The complexity of contaminant issues as they impact wildlife populations is illustrated in the problem of alligator deaths and low egg hatching success at Lake Griffin, Florida. Poor water quality as a result of runoff harboring pesticides and fertilizer changed the composition of plants and animals in the lake. Alligators there switched to a diet mostly of gizzard shad. Gizzard shad contain an enzyme that contributes to thiamine deficiency in alligators, a rational explanation for the mortality.

In Louisiana, where large populations of invasive nutria raze imperiled wetlands, zinc phosphide is used to poison the destructive rodents. Concern of the program's impact on alligators that feed heavily on nutria resulted in research to assess the risk. The conclusion of the study was that the jeopardy to alligators is small. Hopefully, it is outweighed by the ecosystem benefits of removing nutria. It is complicated out there.

Another example involves alligators at Lake Apopka, Florida, situated near a chemical plant that once produced and spilled pesticides. Years of work by Dr. Lou Guillette and his colleagues at the University of Florida have shown that this population also experienced poor hatching success, and male alligators there were found to have abnormally small genitalia attributed to exposure to DDE. The contaminants caused abnormalities in the alligators' endocrine systems in a manner that mimics the hormone estrogen. In essence, the male alligators became more female-like. The Lake Apopka calamity became an example of how contaminants can harm organisms even at levels well below those considered acutely toxic and in unanticipated ways.

Similar endocrine disruptors are not uncommon in human food webs, thus public health concerns were heightened when doctors began noticing related trends in humans. Many scientists now link the increases in premature female puberty, low sperm counts, and testicular cancer to the widespread presence of chemicals in our environment in recent generations. Alligators, then, have become a sentinel species, like coal mine canaries. It is no stretch of the imagination to state that the health of American alligators as it pertains to environmental pollution reflects the same of humans.

Afterword

American alligators, as we know them, have existed for 65 million years or so. As a species, we humans have shared common ecosystems with American alligators for no more than 13,000 years, and most of what we learned of them in that time has been forgotten. Yet within the brief era of recorded history,

Alligator and invasive python in the Everglades. Photo by Lori Oberhofer, U.S. National Park Service.

humans have accumulated a large body of information pertaining to alligators. As a method of assessing its significance and the permeation of alligators into all aspects of our modern culture, consider that in the year 2012 a Google search of the word "alligator" yielded 57,600,000 results in 0.18 seconds. How is this to be interpreted? Which of the results are germane? Is it the article that describes how President Herbert Hoover's son, Allan, allowed his two pet alligators to wander around inside the White House? Does the journal report of a medical discovery that alligator blood products may successfully treat diabetics pass muster? Is it the website for the Alligator Warrior Festival in Lake City, Florida? Or the scientific paper that assesses the impacts of sea level rise on coastal alligators? Assuredly, only those search results that expose a connection between alligators and people should be considered important. That would leave all 57,600,000 as a measure of their standing. And so the fascination continues.

Acknowledgments

Having enjoyed a fulfilling career as a wildlife biologist, I am well aware that most of what we know about wild animals is a result of dedicated, unheralded professionals who spend long days and nights probing about in the rawness of the natural world. In that realm I would like to thank the following researchers and resource managers for providing information for this book: Richard Beagles, Oklahoma Department of Wildlife Conservation; Jeffrey C. Beane, North Carolina State Museum of Natural Sciences; Dr. John Carr, University of Louisiana at Monroe (ULM); Amos Cooper, Texas Parks and Wildlife Department; Dr. Neil Douglas, ULM; Dr. Ruth Elsey, Louisiana Department of Wildlife and Fisheries (LDWF); Ricky Flynt, Mississippi Wildlife, Fisheries, and Parks; Keith Gauldin, Alabama Department of Conservation and Natural Resources; Jeff Hall, North Carolina Wildlife Resources Commission; Glenn A. Harris, United States Fish and Wildlife Service (USFWS); Tom Hess, LDWF; Kelly Irwin, Arkansas Game and Fish Commission; Noel Kinler, LDWF; Dr. Jessica Maisano, University of Texas at Austin; Dr. William Marquardt, Florida Museum of Natural History; Dr. Mark Merchant, McNeese State University; Dr. Steve Reagan, USFWS; Kash Schriefer, USFWS; Dr. Will Selman, LDWF; Derrell Shipes, South Carolina Department of Natural Resources; Dr. Michael Steinberg, University of Alabama; Wayne Syron, USFWS; Greg Waters, Georgia Department of Natural Resources; Allan R. Woodward, Florida Fish and Wildlife Conservation Commission, who provided a most helpful review of the entire manuscript; and especially my friend Paul Yakupzack, USFWS, for logistical support.

I would also like to thank the following individuals for various types of assistance provided to this project: Doug Caldwell, Bryan Craighead, Jeff Donald, Berlin Heck, Robert Rickett, and Ann B. Smith. At University Press of Florida, Senior Acquisitions Editor Sian Hunter shepherded this project and has been a pleasure to work with. Others at the press who have been most helpful are Marthe Walters, Dennis Lloyd, Larry Leshan, Sonia Dickey, Ale Gasso, Olivia Gonzalez, and Kim Lake. Elizabeth Detwiler was my wonderful copy editor. Burg Ransom, nature photographer par excellence, has most graciously allowed me the use of his remarkable work. And every day my wife, Amy, provides nurturing and sustenance that allows me to carry on.

Bibliographic Essay

Introduction

Field diary passages are from my personal diaries. Passage one is from a field diary entry dated June 6, 1981; passage two is from an entry dated September 19, 1983.

Chapter 1. In the Beginning

Evolution/Paleontology

An online map depicting the Earth during the Late Cretaceous period can be found from C. R. Scotese, "Paleomap Project image," accessed July 26, 2012, http://www.scotese.com/images/094.jpg.

General information on archosaurs and crocodilian evolution can be found in the following sources: International Union for Conservation of Nature (IUCN), Crocodile Specialist Group, "Evolution," accessed July 26, 2012, http://www.iucncsg.org/pages/Evolution.html; San Diego Natural History Museum, "After the Dinosaurs—When Crocodiles Ruled," accessed July 26, 2012, http://www.sdnhm.org/archive/exhibits/crocs/tguide/tgcrocs.html; Bob Strauss, "Prehistoric Crocodiles—The Ancient Cousins of the Dinosaurs," accessed July 26, 2012, http://dinosaurs.about.com/od/typesofdinosaurs/a/crocodilians.htm; Enchanted Learning, "Alligator Fossils and Evolution," accessed July 26, 2012, http://www.enchantedlearning.com/subjects/Alligator.

shtml; and GeoChemBio, "*Alligator mississippiensis*," accessed July 26, 2012, http://www.geochembio.com/biology/organisms/alligator/.

The history of "shieldcroc" comes from Univ. of Missouri News Bureau, "New Species of Ancient Crocodile, Ancestor of Today's Species, Discovered by MU Researcher," accessed July 26, 2012, http://munews.missouri.edu/news-releases/2012/0131-new-species-of-ancient-crocodile-ancestor-of-today%E2%80%99s-species-discovered-by-mu-researcher/; and National Geographic Daily News, "Prehistoric 'Shield'-Headed Croc Found," accessed July 26, 2012, http://news.nationalgeographic.com/news/2011/11/111109-shieldcroc-crocodiles-fossils-science-aegisuchus-witmeri/.

For more on the mitochondrial genome, see A. Janke and U. Arnason, "The complete mitochondrial genome of *Alligator mississippiensis* and the separation between recent archosauria (birds and crocodiles)," *Molecular Biology and Evolution* 14, no. 12 (1997): 1266–72.

Brachychampsa montana information is from K. Carpenter and D. Lindsey, "The Dentary of *Brachychampsa montana* Gilmore (Alligatorinae; Crocodylidae), a Late Cretaceous Turtle-Eating Alligator," *Journal of Paleontology* 54, no. 6 (1980): 1213–17.

Chapter 2. Names and Places

Taxonomy

Many sources cite the origin of the word "alligator," including E. A. McIlhenny in *The Alligator's Life History* (Boston: Christopher Publishing House, 1935), 15.

General information on Carl Linnaeus and Linnean taxonomy can be found at Univ. of California Museum of Paleontology, "Carl Linnaeus," accessed July 26, 2012, http://www.ucmp.berkeley.edu/history/linnaeus.html; and *Palaeos Encyclopedia*, "Linnean Taxonomy," accessed July 26, 2012, http://www.palaeos.org/Linnean_taxonomy.

Information on general crocodilian taxonomy can be found in F. W. King and R. L. Burke, *Crocodilian, Tuatara, and Turtle Species of the World: A Taxonomic and Geographical Reference* (Lawrence, Kans.: Assoc. of Systematics Collections, 1989).

Information specifically on alligator taxonomy can be found at the Integrated Taxonomic Information System, "*Alligator mississippiensis*," accessed July 26, 2012, http://www.itis.gov/servlet/SingleRpt/SingleRpt?search_topic=TSN&search_value=551771.

Range, Habitat, and Abundance

Range, habitat, and population data for alligators in North Carolina came from J. G. Hall, North Carolina Wildlife Resources Commission, personal communication, March 22, 2012; and Davidson College Herpetology Lab (2), "American Alligator," accessed July 27, 2012, http://www.herpsofnc.org/herps_of_NC/crocodilians/Allmis/All_mis.html; and "North Carolina Wildlife Profiles—American Alligator," accessed July 27, 2012, http://www.bio.davidson.edu/outreach/NCWRC%20species%20profiles/Reptiles/alligator american.pdf.

Sources in South Carolina include the South Carolina Dept. of Natural Resources (2), "American Alligator," accessed July 27, 2012, http://www.dnr.sc.gov/marine/mrri/acechar/specgal/gator.htm; and "Public Alligator Hunting Season Report 2011," accessed July 27, 2012, http://www.dnr.sc.gov/wildlife/alligator/pdf/huntingreport2011.pdf.

Georgia information can be found at the Savannah River Ecology Laboratory, "American Alligator (*Alligator mississippiensis*)," accessed July 27, 2012, http://srelherp.uga.edu/alligators/allmis.htm; and Georgia Dept. of Natural Resources, "Alligator Fact Sheet," accessed July 27, 2012, http://www.georgia wildlife.com/node/270.

Florida sources include A. R. Woodward, Florida Fish and Wildlife Conservation Commission (FWC), personal communication, March 8, 2012; T. C. Hines, "The Past and Present Status of the Alligator in Florida," *Proc. Annual Conf. Southeastern Assoc. Fish and Wildlife Agencies* 33 (1979): 224–32; and FWC, "How many alligators are in Florida?" accessed July 27, 2012, http://myfwc.custhelp.com/app/answers/detail/a_id/2581/kw/gator/session/L3Rpb WUvMTMzMjM0MjQ2OS9zaWQvc3RpT0xFVGs%3D.

Alabama data were gathered from W. K. Gauldin, Alabama Dept. of Conservation and Natural Resources, personal communication, March 21, 2012 and April 17, 2012; and Auburn Univ. School of Forestry and Wildlife Sciences, "Alabama Wildlife Damage Management—Alligator," accessed July 27, 2012, http://www.aces.edu/forestry/awdm/reptiles/alligator.php.

In Mississippi sources were R. Flynt, Mississippi Dept. of Wildlife, Fisheries and Parks, personal communication, February 28, 2012; and Outdoors Network, "Article 515," accessed July 27, 2012, http://www.outdoors.net/Outdoors/Article/515.

Louisiana sources were the Louisiana Alligator Advisory Council, "*Alligator mississippiensis,*" accessed July 27, 2012, http://www.alligatorfur.com/alligator/alligator.htm; and Restore or Retreat, "Coastal Erosion: Facts &

Figures," accessed July 27, 2012, http://www.restoreorretreat.org/la_erosion_facts.php.

Texas information came from A. Cooper, Texas Parks and Wildlife Dept., personal communication, March 26, 2012; and Texas Parks and Wildlife Dept., "Distribution of American Alligators in Texas," accessed July 31, 2012, http://www.tpwd.state.tx.us/publications/pwdpubs/media/pwd_lf_w7000_0162.pdf.

Arkansas sources were K. J. Irwin, Arkansas Game and Fish Commission, personal communication, February 8, 2012; and C. L. Watt, P. A. Tappe, and M. F. Roth, "Concentrations of American Alligator Populations in Arkansas," *Journal of the Arkansas Academy of Science* 56 (2002): 243–49.

In Oklahoma sources were R. Beagles, Oklahoma Dept. of Wildlife Conservation, personal communication, March 22, 2012; B. Heck, "The American Alligator (*Alligator mississippiensis*) in Oklahoma with Two New Early Records," *Proc. Okla. Acad. Sci.* 86 (2006): 17–21; and Oklahoma Dept. of Wildlife Conservation, "American Alligator," accessed July 27, 2012, http://www.wildlifedepartment.com/wildlifemgmt/species/alligator.htm.

Information on the presence of alligators outside their normal range can be found at the United States Geological Survey (2), "NAS—*Alligator mississippiensis*," accessed July 27, 2012, http://nas.er.usgs.gov/queries/factsheet.aspx?SpeciesID=221; and "NAS—Collection Info," accessed July 27, 2012, http://nas.er.usgs.gov/queries/CollectionInfo.aspx?SpeciesID=221.

Chapter 3. "Their Toes Are Five in Number"

Description

The title of chapter 3 is from a description by E. T. Bennett, W. Harvey, A. R. Branston, and G. T. Wright, *The Tower Menagerie: Comprising the Natural History of the Animals Contained in That Establishment; with Anecdotes of Their Characters and History* (London: Robert Jennings, 1829), 231–32.

A general description of alligators and how they differ from related crocodilians can be found at the following sites: K. P. Schmidt, *The American Alligator*, Chicago Field Museum of Natural History, Leaflet no. 3 (1923), 1–13; International Union for Conservation of Nature, "Crocodiles, Alligators or Gharials?" accessed July 27, 2012, http://www.iucncsg.org/pages/Crocodiles%2C-Alligators-or-Gharials%3F.html; and E. A. McIlhenny, *The Alligator's Life History*, 15–25.

Sources that discuss the characteristics of alligator skin include: B. K. Hall, "Development and Evolution of the Exo-skeleton of the American Alligator," Palaeontological Association, Hall Lab Newsletter 59, accessed August 1,

2012, http://www.palass.org/modules.php?name=palaeo&sec=newsletter&page=90; C. A. Ross and C. D. Roberts, "Scalation of the American Alligator," *FASEB Journal* 7, no. 225 (1979): 1–8; and J. D. Murray, D. C. Deeming, and M. W. Ferguson, "Size-Dependent Pigmentation-Pattern Formation in Embryos of *Alligator mississippiensis*: Time of Initiation of Pattern Generation Mechanism," *Proc. Royal Society of London B*, 239, no. 1296 (1990): 279–93.

Albino and leucistic alligators are discussed in C. C. Lockwood's *The Alligator Book* (Baton Rouge: Louisiana State Univ. Press, 2002), 101–5; and Audubon Nature Institute's "Ghosts of the Wetlands," accessed July 28, 2012, http://www.auduboninstitute.org/blogs/2011/04/ghosts-wetlands. The source of the early alligator quote about the albino alligator in the New York Zoological Park is R. L. Ditmars, *The Reptile Book: A Comprehensive Popularised Work on the Structure and Habits of the Turtles, Tortoises, Crocodilians, Lizards and Snakes Which Inhabit the United States and Northern Mexico* (New York: Doubleday, 1915), 84. The quote about the albino alligators at the National Aquarium is from the National Aquarium, "Oleander, the Albino Alligator, Extends Stay at National Aquarium," accessed February 3, 2012, http://news.aqua.org/2012/02/03/oleander-the-albino-alligator-extends-stay-at-national-aquarium/.

Information on alligator teeth is found at Crocodilian.com, "*Alligator mississippiensis*," accessed July 31, 2012, http://crocodilian.com/cnhc/csp_amis.htm; and National Zoological Park, "Fact Sheets—American Alligator," accessed July 28, 2012, http://nationalzoo.si.edu/Animals/ReptilesAmphibians/Facts/FactSheets/Americanalligator.cfm.

The crushing power of alligator jaws is referenced in B. Dowler, "Contributions to the Natural History of the Alligator," *New Orleans Medical and Surgical Journal* 3 (1846): 311–36; and K. A. Vliet, "Scientists ID Strongest Animal Bite," accessed July 28, 2012, http://animal.discovery.com/news/briefs/20030915/alligator.html. Information on crocodilian palatal valves and salt glands can be found at crocodilian.com, "Palatal Valve," accessed July 28, 2012, http://crocodilian.com/cnhc/cbd-gb2.htm; and "Crocodilian Biology Database," accessed July 28, 2012, http://crocodilian.com/cnhc/cbd-faq-q1.htm.

Gastroliths and matters of alligator digestive tracts are mentioned in International Union for the Conservation of Nature, "The Crocodilian Body," accessed July 30, 2012, http://www.iucncsg.org/pages/The-Crocodilian-Body.html; B. S. Barton, "Facts Relative to the Torpid State of the North American Alligator," *Philosophical Magazine* 23 (1806): 143–45; C. G. Farmer, T. J. Uriona, D. B. Olsen, M. Steenblik, and K. Sanders, "The Right-to-Left Shunt of Crocodilians Serves Digestion," *Physiological and Biochemical Zoology* 81, no. 2 (2008): 125–37; and E. A. McIlhenny, *The Alligator's Life History*, 52.

The circulatory and respiratory systems of alligators are discussed in International Union for the Conservation of Nature, "The Crocodilian Body," accessed July 30, 2012, http://www.iucncsg.org/pages/The-Crocodilian-Body. html; and C. G. Farmer and K. Sanders, "Unidirectional Airflow in the Lungs of Alligators," *Science* 327, no. 5963 (2010): 338–40.

Alligators have a remarkable central nervous system that coordinates an equally amazing array of sensory organs. These topics are discussed in the following sources: A. M. Reese, *The Alligator and Its Allies* (New York: G. P. Putnam's Sons, 1915), 132–35; Statistics Online Computational Resource, "Brain to Body Weight Dataset," accessed July 28, 2012, http://wiki.stat.ucla. edu/socr/index.php/SOCR_Data_Brain2BodyWeight; D. Soares, "An Ancient Sensory Organ in Crocodilians," *Nature* 417 (2002): 241–42; D. Higgs, E. Brittain-Powell, D. Soares, M. Souza, and C. Carr, "Amphibious Auditory Responses of the American Alligator (*Alligator mississipiensis*)," *Journal of Comparative Physiology A: Neuroethology, Sensory, Neural, and Behavioral Physiology* 188, no. 3 (2002): 217–23; F. A. Beach, "Responses of Captive Alligators to Auditory Stimulation," *American Naturalist* 78, no. 779 (1944): 481–505; C. E. Carr, D. Soares, J. Smolders, and J. Z. Simon,, "Detection of Interaural Time Differences in the Alligator," *Journal of Neuroscience* 29, no. 25 (2009): 7978–90; A. Hansen, "Olfactory and Solitary Chemosensory Cells: Two Different Chemosensory Systems in the Nasal Cavity of the American Alligator, *Alligator mississippiensis*," *BMC Neuroscience* 8 (2007): 64; N. Kinler, "Alligators: Natural History Notes," accessed July 30, 2012, http://www. americaswetlandresources.com/wildlife_ecology/plants_animals_ecology/ animals/reptiles/alligator.html; P. B. Johnsen and J. L. Wellington, "Detection of Glandular Secretions by Yearling Alligators," *Copeia* 1982, no. 3 (1982): 705–8; C. Gans and D. Crews, "Hormones, Brain, and Behavior," in *Biology of the Reptilia—Volume 18* (Chicago: Univ. of Chicago Press, 1992), 203; International Union for the Conservation of Nature, "The Crocodilian Body," accessed July 30, 2012, http://www.iucncsg.org/pages/The-Crocodilian-Body. html; A. J. Sillman, S. J. Ronan, and E. K. Loew, "Histology and Microspectrophotometry of the Photoreceptors of a Crocodilian, *Alligator mississippiensis*, *Proc. Royal Society of London B* 243, no. 1306 (1991): 93–98; and P. A. Murphy, "Celestial Compass Orientation in Juvenile American Alligators (*Alligator mississippiensis*)," *Copeia* 1981, no. 3 (1981): 638–45.

Locomotion

Sources on alligator locomotion include: International Union for the Conservation of Nature, "Locomotion," accessed July 30, 2012, http://www.iucncsg. org/pages/Locomotion.html; F. E. Fish, "Kinematics of Undulatory Swimming in the American Alligator," *Copeia* 1984, no. 4 (1984): 839–43; and S. M. Reilly and J. A. Elias, "Locomotion in *Alligator mississippiensis*: Kinematic Effects of Speed and Posture and Their Relevance to the Sprawling-to-Erect Paradigm," *Journal of Experimental Biology* 201 (1998): 2559–74.

Temperature Regulation

Alligators are "cold-blooded." This physiological trait and its behavioral consequences are discussed further at International Union for the Conservation of Nature, "Temperature Regulation," accessed July 30, 2012, http://www.iucncsg. org/pages/Temperature-Regulation.html; H. F. Percival et al., "Thermoregulation of the American Alligator in the Everglades," accessed July 30, 2012, http://fl.biology.usgs.gov/posters/Everglades/Alligator_Thermoregulation/ alligator_thermoregulation.html; N. Orobello, "Thermoregulation and its Effects in the American Alligator (*Alligator mississippiensis*)," accessed July 30, 2012, http://www.bio.davidson.edu/people/midorcas/animalphysiology/ websites/2011/Orobello/Index.htm; and T. Joanen and L. McNease, "The Effects of a Severe Winter Freeze on Wild Alligators in Louisiana," *Proc. IUCN, Crocodile Specialist Group, Papua, New Guinea*, 1988.

Chapter 4. To Build an Alligator

Growth and Size

The source of the 1893 government report on alligator growth is H. M. Smith, "Of Alligators and Oysters," in *Fisheries of the South Atlantic States* (Washington, D.C.: GPO, 1893). Other misconceptions are recorded in South Carolina Parks, "Common Myths and the Truth about Alligators," accessed July 30, 2012, http://www.southcarolinaparks.com/files/State%20Parks/Wildlife%20 Page/WW_alligators.pdf. More accurate data on the subject can be found at E. A. McIlhenny, *The Alligator's Life History*, 57–65; R. H. Chabreck and T. Joanen, "Growth Rates of American Alligators in Louisiana," *Herpetologica* 35, no. 1 (1979): 51–57; R. M. Elsey, T. Joanen, L. McNease, and N. Kinler, "Growth Rates and Body Condition Factors of *Alligator mississippiensis* in Coastal Louisiana Wetlands: A Comparison of Wild and Farm-Released Juveniles," *Comp. Biochem. Physiol. A* 103, no. 4 (1992): 667–72; and W. L. Rootes, R. H. Chabreck, V. L. Wright, B. W. Brown, and J. J. Hess, "Growth Rates of American Alligators in Estuarine and Palustrine Wetlands in Louisiana," *Estuaries* 14, no. 4 (1991): 489–94.

Information regarding Edward McIlhenny's giant alligator can be found at: E. A. McIlhenny, *The Alligator's Life History*, 60–61; McIlhenny Company,

"Some Common Myths," accessed July 30, 2012, http://www.tabasco.com/mcilhenny-company/faqs-archives/myths/; R. M. Grace, "Edward A. McIlhenny: Businessman, Naturalist, Author . . . Fibber," *Metropolitan News-Enterprise*, accessed July 31, 2012, http://www.metnews.com/articles/2004/reminiscing102104.htm.

Alligator aging techniques and other materials related to very large alligators are discussed at: A. R. Woodward, J. H. White, and S. B. Linda, "Maximum Size of the Alligator (*Alligator mississippiensis*)," *Journal of Herpetology* 29, no. 4 (1995): 507–13; N. M. Scott, M. F. Haussmann, R. M. Elsey, P. L. Trosclair III, and C. M. Vleck, "Telomere Length Shortens with Body Length in *Alligator mississippiensis*," *Southeastern Naturalist* 5, no. 4 (2006): 685–92; Pittsburgh Zoo, "American Alligator," accessed July 30, 2012, http://www.pittsburghzoo.org/animal.aspx?id=72; A. M. Reese, *The Alligator and Its Allies*, 16; and R. Flynt, Mississippi Dept. of Wildlife, Fisheries, and Parks, personal communication. Data for the Lacassine NWR studies are found in the refuge files and include Annual Narrative Reports (1981–2011) and Harvest Data Reports (1983–2011).

Sources of modern alligator records in table 3 follow:

Alabama: al.com, "Huge alligator killed in west-central Alabama: 14 feet, 2 inches, 838 pounds," accessed July 31, 2012, http://www.al.com/sports/index.ssf/2011/08/huge_alligator_killed_in_west-.html.

Arkansas: Arkansas Outdoors, "Arkansas Alligator Hunting," accessed July 31, 2012, http://www.arkansasoutdoorsonline.com/arkansas-alligator-hunting/.

Florida: FWC, "Alligator Facts," accessed July 31, 2012, http://myfwc.com/wildlifehabitats/managed/alligator/facts/.

Georgia: Field and Stream, "Three Gigantic Gators Killed in One Week in South Carolina and Georgia," accessed July 31, 2012, http://www.fieldandstream.com/photos/gallery/hunting/2010/09/three-gigantic-gators-killed-one-week-south-carolina-georgia?photo=8%23node-1001370425.

Louisiana: Woodward, White, and Linda, "Maximum Size of the Alligator (*Alligator mississippiensis*)."

Mississippi: Mississippi Sportsman, "Foursome take state-record alligator," accessed July 31, 2012, http://www.ms-sportsman.com/reader.php?id=1178.

North Carolina: J. G. Hall, North Carolina Wildlife Resources Commission, personal communication, March 8, 2012.

South Carolina: CBS News, "Five-Foot Woman Kills 13-Foot Alligator," accessed July 31, 2012, http://www.cbsnews.com/stories/2010/09/18/earlyshow/saturday/main6879091.shtml.

Texas: Texas Parks and Wildlife Dept., "Distribution of American Alliga-

tors in Texas," accessed July 31, 2012, http://www.tpwd.state.tx.us/publications/
pwdpubs/media/pwd_lf_w7000_0162.pdf.

Food Habits

Studies that pertain to alligator food habits include: M. F. Delany and C. L. Abercrombie, "American Alligator Food Habits in Northcentral Florida," *Journal of Wildlife Management* 50, no. 2 (1986): 348–53; M. J. Fogarty and J. D. Albury, "Late Summer Foods of Young Alligators in Florida," *Proc. 21st Annual Conf. of Southeastern Assoc. of Game and Fish Commissioners* (1967), 220–22; M. F. Delany, A. R. Woodward, and I. H. Kochel, "Nuisance Alligator Food Habits in Florida," *Florida Field Naturalist* 16 (1988): 90–96; L. McNease and T. Joanen, "Alligator Diets in Relation to Marsh Salinity," *Proc. Annual Conf. Southeastern Assoc. Fish and Wildlife Agencies* 31 (1977): 36–40; R. M. Elsey, L. McNease, T. Joanen, and N. Kinler, "Food Habits of Native Wild and Farm-released Juvenile Alligators," *Proc. Annual Conf. Southeastern Assoc. Fish and Wildlife Agencies* 46 (1992): 57–66; D. Taylor, "Fall Foods of Adult Alligators from Cypress Lake Habitat, Louisiana," *Proc. Annual Conf. Southeastern Assoc. Fish and Wildlife Agencies* 40 (1986): 338–41; A. N. Rice, "Diet and Condition of American Alligators in Three Central Florida Lakes," MS thesis, Univ. of Florida, 1–100, 2004; M. F. Delany, S. B. Linda, and C. T. Moore, "Diet and Condition of American Alligators in 4 Florida Lakes," *Proc. Annual Conf. Southeastern Assoc. Fish and Wildlife Agencies* 53 (1999): 375–89; S. W. Gabrey, "Demographic and Geographic Variation in Food Habits of American Alligators (*Alligator mississippiensis*) in Louisiana," *Herpetological Conservation and Biology* 5, no. 2 (2010): 241–50; D. T. Saalfeld, W. C. Conway, and G. E. Calkins, "Food Habits of American Alligators (*Alligator mississippiensis*) in East Texas," *Southeastern Naturalist* 10, no. 4 (2011): 659–72.

References to accounts of alligators eating non-native species include: E. A. McIlhenny, *The Alligator's Life History*, 55; F. A. Ober, "Florida and the West Indies," *Journal of the American Geographical Society of New York* 18 (1886): 183–214; M. Reid, "Osceolo, The Seminole," *Graham's Illustrated Magazine* 53 (1858): 226–31; *Tallahassee Sentinel*, February 19, 1870, "An Elephant and Two Camels Attacked by Alligators."

Other feeding behavior is described by L. Pierrard and M. G. Frick, "Scavenging of Turtle Carcasses by American Alligators, *Alligator mississippiensis*, in Georgia, USA," *Proc. 20th Annual Sea Turtle Symposium* (2000), 212–13; M. F. Delany, "What Do Alligators Eat?" *Florida Wildlife* (Nov.-Dec. 1987): 7–8; W. L. Rootes and R. H. Chabreck, "Cannibalism in the American Alligator," *Herpetologica* 49, no. 1 (1993): 99–107; M. F. Delany, A. R. Woodward, R. A. Kiltie, and C. T. Moore, "Mortality of American Alligators Attributed to Cannibalism," *Herpetologica* 67, no. 2 (2011): 174–85; L. A. Hayes-Odum and D. Jones, "Effects of Drought on American Alligators (*Alligator mississippiensis*) in Texas," *Texas Journal of Science* 45, no. 2 (1993): 1–3; and GeoChemBio, "*Alligator mississippiensis,*" accessed July 26, 2012, http://www.geochembio.com/biology/organisms/alligator/.

The information in table 4 is derived from sources listed for this subchapter.

Reproduction

Alligator courtship, mating, and nest site sources include K. A. Vliet, "Social Displays of the American Alligator (*Alligator mississippiensis*)," *American Zoologist* 29 (1989): 1019–31; T. Joanen, "Nesting Ecology of Alligators in Louisiana," *Proc. Southeastern Assoc. Game and Fish Commissioners Conf.* 23 (1969): 141–51; T. Joanen and L. McNease, "Time of Egg Deposition for the American Alligator," *Proc. Annual Conf. Southeastern Assoc. Fish and Wildlife. Agencies* 33 (1979): 15–19; D. H. Gist, A. Bagwill, V. Lance, D. M. Sever, and R. M. Elsey, "Sperm Storage in the Oviduct of the American Alligator," *Journal of Experimental Zoology A* 309 (2008): 581–87; L. M. Davis, T. C. Glenn, R. M. Elsey, H. C. Dessauers, and R. H. Sawyer, "Multiple Paternity and Mating Patterns in the American Alligator, *Alligator mississippiensis,*" *Molecular Ecology* 10 (2001): 1011–24; S. L. Lance, T. D. Tuberville, L. Dueck, C. Holz-Schietinger, P. L. Trosclair III, R. M. Elsey, and T. C. Glenn, "Multiyear Multiple Paternity and Mate Fidelity in the American Alligator, *Alligator mississippiensis,*" *Molecular Ecology* 18 (2009): 4508–20; V. A. Lance, "Alligator Physiology and Life History: The Importance of Temperature," *Experimental Gerontology* 38, no. 7 (2003): 801–5; A. R. Woodward, T. Hines, C. Abercrombie, and C. Hope, "Spacing Patterns in Alligator Nests," *Journal of Herpetology* 18, no. 1 (1984): 8–12; L. McNease, N. Kinler, T. Joanen, David Richard, and Darren Richard, "Distribution and Relative Abundance of Alligator Nests in Louisiana Coastal Marshes," *Proc. IUCN, Crocodile Specialist Group Mtg.*, May 2–6, 1994; R. M. Elsey, P. L. Trosclair III, and T. C. Glenn, "Nest-site Fidelity in American Alligators in a Louisiana Coastal Marsh," *Southeastern Naturalist* 7, no. 4 (2008): 737–43; and D. T. Saalfeld, "American Alligator (*Alligator mississippiensis*) Ecology in Inland Wetlands of East Texas," PhD Diss., Stephen F. Austin State Univ., 2010.

More information on alligator nest construction, egg-laying, clutch size, and incubation can be found at: T. Joanen and L. McNease, "Ecology and Physiology of Nesting and Early Development of the American Alligator," *American Zoologist* 29, no. 3 (1989): 987–98; T. Joanen and L. McNease, "Reproductive Biology of the American Alligator in Southwest Louisiana," *Proc.*

SSAR Symposium on Reproductive Biology and Diseases of Captive Reptiles 1 (1980): 153–59; T. Joanen and L. McNease, "Nesting Chronology of the American Alligator and Factors Affecting Nesting in Louisiana," *Proc. First Annual Alligator Production Conf., Gainesville, Fla., Feb. 12–13, 1981,* 1–14; T. M. Goodwin and W. R. Marion, "Aspects of the Nesting Ecology of American Alligators (*Alligator mississippiensis*) in North-Central Florida," *Herpetologica* 34 (1978): 43–47; C. S. Wink and R. M. Elsey, "Changes in Femoral Morphology During Egg-laying in *Alligator mississippiensis,*" *Journal of Morphology* 189 (1986): 183–88; and M. W. Ferguson, "The Structure and Composition of the Eggshell and Embryonic Membranes of *Alligator mississippiensis,*" *Transactions of the Zoological Society of London* 36 (1982): 99–152.

Sources for the natural impediments to successful nesting include R. M. Elsey, "The Effects of Wildfires on Alligator Nests on Rockefeller Refuge," *Proc. Annual Conf. of the Southeastern Assoc. of Fish and Wildlife Agencies* 50 (1996): 532–40; R. M. Elsey and E. B. Moser, "The Effect of Lightning Fires on Hatchability of Alligator Eggs," *Herpetological Natural History* 9, no. 1 (2002): 51–56; T. Joanen, L. McNease, and G. Perry, "Effects of Simulated Flooding on Alligator Eggs," *Proc. Annual Conf. Southeastern Assoc. Fish and Wildlife Agencies* 31 (1977): 33–35; D. C. Deitz and T. C. Hines, "Alligator Nesting in North-Central Florida," *Copeia* 1980, no. 2 (1980): 249–58; and Joanen, "Nesting Ecology of Alligators in Louisiana."

The fascinating process of temperature-dependent sex determination is detailed in M. W. Ferguson and T. Joanen, "Temperature of Egg Incubation Determines Sex in *Alligator mississippiensis,*" *Nature* 29 (1982): 850–53; M. W. Ferguson and T. Joanen, "Temperature-dependent Sex Determination in *Alligator mississippiensis,*" *Journal of Zoology of London* 200, no. 2 (1983): 143–47; and V. A. Lance, R. M. Elsey, and J. W. Lang, "Sex Ratios of American Alligators (Crocodylidae): Male or Female Biased?," *Journal of Zoology of London* 252 (2000): 71–78.

Chapter 5. Root Hog, or Die

Movement

Sources that describe alligator movements include: T. Joanen and L. McNease, "A Telemetric Study of Nesting Female Alligators on Rockefeller Refuge, Louisiana," *Proc. Southeastern Assoc. Game and Fish Commissioners Conf.* 24 (1970): 175–93; R. H. Chabreck, "The Movement of Alligators in Louisiana," *Proc. Southeastern Assoc. Game and Fish Commissioners Conf.* 19 (1965): 102–10; T. M. Goodwin and W. R. Marion, "Seasonal Activity Ranges and Habitat Preferences of Adult Alligators in a North-Central Florida Lake," *Journal of Herpetology* 13, no. 2 (1979): 157–64; T. Joanen and L. McNease, "A Telemetric Study of Adult Male Alligators on Rockefeller Refuge, Louisiana," *Proc. Southeastern Assoc. Game and Fish Commissioners Conf.* 26 (1972): 252–75; V. A. Lance, R. M. Elsey, P. E. Trosclair III, and L. A. Nunez, "Long-distance Movement by American Alligators in Southwest Louisiana," *Southeastern Naturalist* 10, no. 3 (2011): 389–98; R. M. Elsey, V. A. Lance, and P. L. Trosclair III, "Evidence for Long-Distance Migration by Wild American Alligators," in *Proc. 17th Working Meeting of the Crocodile Specialist Group. Darwin, Australia,* 2004; R. M. Elsey, "Unusual Offshore Occurrence of an American Alligator," *Southeastern Naturalist* 4, no. 3 (2005): 533–36; and R. M. Elsey and C. Aldrich, "Long-distance Displacement of a Juvenile Alligator by Hurricane Ike," *Southeastern Naturalist* 8, no. 4 (2009): 746–49.

Mortality Factors

Some of the many life-threatening hazards faced by alligators are discussed in the following references: FWC, "Alligator Facts;" R. H. Hunt and J. J. Ogden, "Selected Aspects of the Nesting Ecology of American Alligators in the Okefenokee Swamp," *Journal of Herpetology* 25, no. 4 (1991): 448–53; C. R. Allen, K. G. Rice, D. P. Wojcik, and H. F. Percival, "Effect of Red Imported Fire Ant Envenomization on Neonatal American Alligators," *Journal of Herpetology* 31, no. 2 (1997): 318–21; R. McBride and C. McBride, "Predation of a Large Alligator by a Florida Panther," *Southeastern Naturalist* 9, no. 4 (2010): 854–56; Delany et al., "Mortality of American Alligators Attributed to Cannibalism;" W. L. Rootes and R. H. Chabreck, "Cannibalism in the American Alligator;" and R. M. Elsey, V. A. Lance, P. L. Trosclair III, and M. Merchant, "Effect of Hurricane Rita and a Severe Drought on Alligators in Southwest Louisiana," in *Proc. 19th Working Meeting of the Crocodile Specialist Group, Santa Cruz, Bolivia,* 2008.

The effects of parasites and diseases on alligators are examined in E. B. Shotts, Jr., J. L. Gaines, Jr., L. Martin, and A. K. Prestwood, "*Aeromonas*-Induced Deaths Among Fish and Reptiles in an Eutrophic Inland Lake," *Journal of the American Veterinary Medical Assoc.* 161, no. 6 (1972): 603–7; J. J. Daly, "Bacteremia Associated with Mortality in an Arkansas Alligator," *Proc. Arkansas Academy of Science* 45 (1991): 121–22; Univ. of Florida News, "Toxic Algae Possible Cause of Increased Gator Deaths in Lake Griffin," accessed August 2, 2012, http://news.ufl.edu/2000/06/21/algae/; R. M. McNew, R. M. Elsey, T. R. Rainwater, E. J. Marsland, and S. M. Presley, "Survey for West Nile Virus Infection in Free-ranging American Alligators in Louisiana," *Southeastern Naturalist* 6, no. 4 (2007): 737–42; E. R. Jacobson,

P. E. Ginn, J. M. Troutman, L. Farina, L. Stark, K. Klenk, K. L. Burkhalter, and N. Komar, "West Nile Virus Infection in Farmed American Alligators (*Alligator mississippiensis*) in Florida," *Journal of Wildlife Diseases* 41, no. 1 (2005): 96–106; T. C. Hazen, J. M. Aho, T. M. Murphy, G. W. Esch, and G. D. Schmidt, "The Parasite Fauna of the American Alligator (*Alligator mississippiensis*) in South Carolina," *Journal of Wildlife Diseases* 14 (1978): 435–39; A. K. Davis, R. V. Hovan III, A. M. Grosse, B. B. Harris, B. S. Metts, D. E. Scott, and T. D. Tuberville, "Gender Differences in Haemogregarine Infections in American Alligators (*Alligator mississippiensis*) at Savannah River, South Carolina, USA," *Journal of Wildlife Diseases* 47, no. 4 (2011): 1047–49; R. H. Cherry and A. L. Ager, Jr., "Parasites of American Alligators (*Alligator mississippiensis*) in South Florida," *Journal of Parasitology* 68, no. 3 (1982): 509–10; and J. C. Nifong and M. G. Frick, "First Record of the American Alligator (*Alligator mississippiensis*) as a Host to the Sea Turtle Barnacle (*Chelonibia testudinaria*)," *Southeastern Naturalist* 10, no. 3 (2011): 557–60. The source of the Herodotus comment is Dowler, "Contributions to the Natural History of the Alligator."

The Keystone Species

Information on the role of alligators as a keystone species in their wetland ecosystems is found in the following sources: United States Geological Survey (2), "American Alligator Ecology and Monitoring for the Comprehensive Everglades Restoration Plan," accessed August 2, 2012, http://pubs.usgs.gov/fs/2004/3105/pdf/fs-2004-3105-Rice.pdf, and "Role of American Alligator (*Alligator mississippiensis*) in Measuring Restoration Success in the Florida Everglades," accessed August 2, 2012, http://sofia.usgs.gov/geer/2003/posters/gator_restore/; Savannah River Ecology Laboratory, "American Alligator," accessed August 2, 2012, http://srelherp.uga.edu/SPARC/PDFs/AlligatorBrochure.pdf; Univ. of Florida—The Croc Docs, "Ecology of Everglades Alligator Holes," accessed August 2, 2012, http://crocdoc.ifas.ufl.edu/publications/posters/ecologyofalligatorholes/; M. R. Campbell and F. J. Mazzotti, "Characterization of Natural and Artificial Alligator Holes," *Southeastern Naturalist* 3, no. 4 (2004): 583–94; C. Bondavalli and R. E. Ulanowicz, "Unexpected Effects of Predators Upon Their Prey: The Case of the American Alligator," *Ecosystems* 2, no. 1 (1999): 49–63; M. L. Palmer and F. J. Mazzotti, "Structure of Everglades Alligator Holes," *Wetlands* 24, no. 1 (2004): 115–22; K. M. Enge , H. F. Percival, K. G. Rice, M. L. Jennings, G. R. Masson, and A. R. Woodward, "Summer Nesting of Turtles in Alligator Nests in Florida," *Journal of Herpetology* 34, no. 4 (2000): 497–503; P. M. Hall and A. J. Meier, "Reproduction and Behavior of Western Mud Snakes

(*Paranoia abacura reinwardtii*) in American Alligator Nests," *Copeia* 1993, no. 1 (1993): 219–22; M. R. Tansey, "Isolation of Thermophilic Fungi from Alligator Nesting Material," *Mycologia* 65, no. 3 (1973): 594–601; and W. H. Leigh, "The Florida Spotted Gar, As the Intermediate Host for *Odhneriotrema incommodum* from *Alligator mississippiensis*," *Journal of Parasitology* 46, no. 5 (1960): 1–16.

Chapter 6. First Encounters

Alligators and the First People

Sources that describe the paleo-environment include: NOAA Paleoclimatology, "Summary of 100,000 Years," accessed August 2, 2012, http://www.ncdc.noaa.gov/paleo/ctl/100k.html; United States Environmental Protection Agency, "The Ice Age (Pleistocene Epoch)," accessed August 2, 2012, http://www.epa.gov/gmpo/edresources/pleistocene.html; Texas Beyond History, "The Prehistory of the Texas Coastal Zone: 10,000 Years of Changing Environment and Culture," accessed August 2, 2012, http://www.texasbeyondhistory.net/coast/prehistory/images/intro.html; Univ. of Florida—IFAS Extension, "Florida's Geological History," accessed August 2, 2012, http://edis.ifas.ufl.edu/uw208.

Information on alligator interactions with prehistoric humans is found in the following sources: Lost Worlds, "Sapelo Shell Rings (2170 BC)," accessed August 2, 2012, http://lostworlds.org/sapelo_shell_rings/; Poverty Point, "Foods," accessed August 2, 2012, http://www.crt.state.la.us/archaeology/virtualbooks/poverpoi/food.htm; Historic Spanish Point, "Prehistory," accessed August 2, 2012, http://historicspanishpoint.org/history/prehistory; *Sun Sentinel*, July 23, 2011, "Signs of Early Settlements Found on Lauderdale Barrier Island," accessed August 2, 2012, http://articles.sun-sentinel.com/2011-07-23/news/fl-tequesta-lauderdale-beach-20110723_1_barrier-island-archaeologists-tequesta-indians; Granger Collection, "Calusa Figurehead," accessed August 2, 2012, http://www.granger.com/results.asp?inline=true&image=0174453&wwwflag=1&imagepos=10&screenwidth=1264; G. Gibbon, ed., *Archaeology of Prehistoric Native America: An Encyclopedia* (New York: Garland Publishing, 1998), 519; R. J. Russell and H. V. Howe, "Cheniers of Southwestern Louisiana," *Geographical Review* 25, no. 3 (1935): 449–61; Ohio Archaeology, "Alligator Mound," accessed August 2, 2012, http://www.ohio-archaeology.blogspot.com/2011/07/alligator-mound.html.

European accounts of alligator/Native American interactions are revealed in the New World, "Florida Indians Gallery—Narrative of Le Moyne," accessed August 3, 2012, http://thenewworld.us/florida-indians-gallery/18/; the New World, "Memoir of Fontaneda (Version 2)," accessed August 3, 2012, http://thenewworld.us/memoir-of-fontaneda-version-2/; R. L. Hall, *An Archaeology of the Soul* (Champaign: Univ. of Illinois Press, 1997), 14; Texas State Historical Association, "Karankawa Indians," accessed August 3, 2012, http://www.tshaonline.org/handbook/online/articles/bmk05; P. Forest (1832), *Voyage aux Etats-Unis de l'Amerique en 1831,* in V. L. Glasgow's *A Social History of the American Alligator* (New York: St. Martin's Press, 1991), 1–2; Native Languages, "Native American Alligator Mythology," accessed August 3, 2012, http://www.native-languages.org/legends-alligator.htm; and United States National Park Service, "Archeology Program—Ancient Architects of the Mississippi," accessed August 3, 2012, http://www.nps.gov/archeology/feature/native.htm.

The New People—Early Years

Some early European encounters with alligators are documented in the following sources: J. Higginbotham, ed., *The Journal of Sauvole* (Mobile, Ala.: Colonial Books, 1969), 26; A. S. Le Page du Pratz, *Histoire de la Louisiana,* vol. 1 (Paris, 1758); and C. A. Brasseaux, H. D. Hoese, and T. C. Michot, "Pioneer Amateur Naturalist Louis Judice," *Louisiana History* 45, no. 1 (2004): 71–103. All of the William Bartram passages are found in his *Travels Through North and South Carolina, Georgia, East and West Florida, the Cherokee Country, etc.* (London, 1794). John J. Audubon writes about alligators in "Observations on the Natural History of the Alligator," *Museum of Foreign Literature and Science* 11 (1827): 272–75. Other relevant materials include: Punch, "Game Alligators," *Littell's Living Age* 10 (1846): 211; M. P. Hendrix, *Down and Dirty: Archaeology of the South Carolina Lowcountry* (Charleston: History Press, 2006), 69; H. Collins interview in *Slave Narratives: A Folk History of Slavery in the United States from Interviews with Former Slaves,* vol. 16—Texas Narratives WPA, Federal Writer's Project, Library of Congress microfilm, 1941; H. Kennedy interview in *Slave Narratives,* vol. 9—Mississippi Narratives; and S. Northup, *Twelve Years A Slave,* edited by S. Eakin and J. Logsdon (Baton Rouge: Louisiana State Univ. Press, 1975), 103.

Information about the role of the alligator in the American Civil War and primary sources for soldier accounts can be found in my *Flora and Fauna of the Civil War: An Environmental Reference Guide* (Baton Rouge: Louisiana State Univ. Press, 2010), 115–18.

Chapter 7. Near Fatal Attraction

Beyond Sustainability

The reckless overharvest of alligators in the late nineteenth and twentieth centuries is evaluated in the following sources: R. H. Chabreck, "The American Alligator—Past, Present and Future," *Proc. Southeastern Assoc. Game and Fish Commissioners Conf.* 21 (1968): 554–58; A. R. Woodward, personal communication, October 29, 2012; G. B. Goode, *The Fisheries and Fishery Industries of the United States* (Washington, D.C.: GPO, 1884), 146; F. Moore, "Madame Sarah Bernhardt's Alligator Hunt," *Wide World Magazine* 10 (Oct. 1902-March 1903): 19–22; R. D. Carson, "List of Additions to the Menagerie During the Year Ending February 28th, 1918," *Annual Report of the Board of Directors of the Zoological Society of Philadelphia* (1918), 18–22; author unknown, "Trapping Alligators and Rattlers—New York Post," *Current Literature*, vol. 27 (1900), 165–66; Ober, "Florida and the West Indies;" author unknown, "Alligator Hunting in Florida—*Chicago Tribune*," *Current Literature, vol. 2* (1889), 430; A. M. Reese, "The Breeding Habits of the Florida Alligator," *Syracuse Univ. Publications—Contributions from the Zoological Library* 3 (1906): 381–87; author unknown, "Shoe Paragraphs," *Shoe Worker's Journal* 7, no. 1 (1906): 21; author unknown, "The Alligator Satchel," *Forest and Stream* 80 (Feb. 15, 1913): 208; S. C. Arthur, *The Fur Animals of Louisiana* (New Orleans: Louisiana Dept. of Conservation, 1931), 64. Data on the number and prices of Florida alligator hides (table 5) are from E. R. Allen and W. T. Neill, "Increasing Abundance of the Alligator in the Eastern Portion of Its Range," *Herpetologica* 5, no. 6 (1949): 109–12.

Sources that depict the status of alligators in the 1960s and 1970s include: Chabreck, "The American Alligator—Past, Present and Future"; A. R. Woodward, D. N. David, and T. C. Hines, "American Alligator Management in Florida," in *Proc. 3rd Southeastern Nongame and Endangered Wildlife Symposium*, edited by R. R. Odom, K. A. Riddleberger, and S. C. Ozier (1987), 98–113; A. R. Woodward, "American Alligators in Florida," in *Our Living Resources: A Report to the Nation on the Distribution, Abundance, and Health of U.S. Plants, Animals and Ecosystems*, edited by E. T. LaRoe, G. S. Farris, C. E. Puckett, P. D. Doran, and M. J. Mac (Washington, D.C.: USDI, National Biological Service, 1995), 127–29; and K. J. Irwin, "American Alligators are Naturals in Arkansas," *Arkansas Wildlife* (May–June 2006): 10–11.

As they pertain to alligators, the Lacey Act and Endangered Species Act are discussed at Animal Legal and Historical Center, "History and Development of the Lacey Act," accessed August 3, 2012, http://www.animallaw.info/

articles/arus16publlr27.htm#2; U.S. Fish and Wildlife Service (2), "Species Profile—American Alligator," accessed August 3, 2012, http://ecos.fws.gov/speciesProfile/profile/speciesProfile.action?spcode=C000, and "Endangered Species—American Alligator," accessed August 3, 2012, http://www.fws.gov/endangered/esa-library/pdf/alligator.pdf; and *Federal Register,* vol. 40, "Proposal to Reclassify the American Alligator," accessed August, 3, 2012, http://ecos.fws.gov/docs/federal_register/fr63.pdf.

Modern Management

Information on the nuisance alligator programs in Florida and Louisiana can be found at: FWC, "Statewide Alligator Nuisance Program," accessed August 4, 2012, http://myfwc.com/wildlifehabitats/managed/alligator/harvest/; Louisiana Dept. of Wildlife and Fisheries (2), "Nuisance Alligators," accessed August 4, 2012, http://www.wlf.louisiana.gov/wildlife/nuisance-alligators, and "Alligator Program," accessed August 4, 2012, http://www.wlf.louisiana.gov/wildlife/alligator-program.

Information on CITES can be found at: Convention on International Trade in Endangered Species of Wild Fauna and Flora (CITES), accessed August 4, 2012, http://www.cites.org/.

Sources for alligator hunting regulations follow by state:

Louisiana: Louisiana Dept. of Wildlife and Fisheries, "Alligator Hunting Regulations Overview," accessed August 4, 2012, http://www.wlf.louisiana.gov/alligator-hunting-regulations-overview.

Florida: FWC, "Statewide Alligator Harvest Program," accessed August 4, 2012, http://myfwc.com/wildlifehabitats/managed/alligator/harvest/.

Texas: Texas Parks and Wildlife Dept., "Managing Alligator Populations," accessed August 4, 2012, http://www.tpwd.state.tx.us/huntwild/wild/species/alligator/management/index.phtml.

South Carolina: South Carolina Dept. of Natural Resources, "Alligators—Hunting," accessed August 4, 2012, http://www.dnr.sc.gov/wildlife/alligator/.

Georgia: Georgia Dept. of Natural Resources, "Alligator Hunting Season for 2012," accessed August 4, 2012, http://www.georgiawildlife.com/node/610.

Mississippi: Mississippi Wildlife, Fisheries, and Parks, "Alligator Program," accessed August 4, 2012, http://www.mdwfp.com/wildlife-hunting/alligator-program.aspx.

Alabama: Alabama Dept. of Conservation and Natural Resources, "Regulations and Application Instructions—Alligator Hunting Season," accessed August 4, 2012, http://outdooralabama.com/hunting/game/alligatorhunthome/regulation.cfm.

Arkansas: Arkansas Game and Fish Commission, "Licenses and Permits—Alligator Hunt Permits," accessed August 4, 2012, http://www.agfc.com/licenses/pages/PermitsSpecialAlligator.aspx.

Sources for the discussion on early alligator farms include: Chameleon, "Alligator Hunting," *Fores Sporting Notes and Sketches* 11 (1894): 107–12; E. W. Mayo, "An Alligator Farm," *Travel* 9, no. 3 (1905): 170–72; and A. Inkersley, "The California Alligator Ranch," *Overland Monthly* 56 (1910): 533–39.

Consulted sources with information pertaining to various aspects of the alligator farming industry including biological concerns are: R. C. Noble, R. McCartney, and M. W. Ferguson, "Lipid and Fatty Acid Compositional Differences between Eggs of Wild and Captive-breeding Alligators (*Alligator mississippiensis*): An Association with Reduced Hatchability?" *Journal of Zoology* 230, no. 4 (2009): 639–40; A. R. Woodward, "Alligator Ranching Research in Florida, USA," in *Wildlife Management: Crocodiles and Alligators*, edited by G. J. Webb, S. C. Manolis, and P. J. Whitehead (Sydney, Australia: Surrey Beatty and Sons in Association with the Conservation Commission of the Northern Territory, 1987), 363–67; R. M. Elsey, T. Joanen, L. McNease, and N. Kinler, "Growth Rates and Body Condition Factors of *Alligator mississippiensis* in Coastal Louisiana Wetlands"; E. R. Jacobson et al., "West Nile Virus Infection in Farmed American Alligators (*Alligator mississippiensis*) in Florida"; T. L. Clippinger, R. A. Bennett, C. M. Johnson, K. A. Vliet, S. L. Deem, J. Oros, E. R. Jacobson, I. M. Schumacher, D. R. Brown, and M. B. Brown, "Morbidity and Mortality Associated with a New Mycoplasma Species from Captive American Alligators (*Alligator mississippiensis*)," *Journal of Zoo and Wildlife Medicine* 31, no. 3 (2000): 303–14; Agricultural Marketing Resource Center, "Alligator Profile," accessed August 4, 2012, http://www.agmrc.org/commodities_products/aquaculture/alligator_profile.cfm; Louisiana Dept. of Wildlife and Fisheries, "General Alligator Information," accessed August 4, 2012, http://www.wlf.louisiana.gov/general-alligator-information; T. J. Lane and K. C. Ruppert, "Alternative Opportunities for Small Farms: Alligator Production Review," Univ. of Florida Cooperative Extension Service, Fact Sheet RF-AC002, 1998; Texas Parks and Wildlife Dept., "Alligator Farming in Texas," accessed August 4, 2012, http://www.tpwd.state.tx.us/publications/pwdpubs/media/pwd_bk_w7000_1433.pdf; A. R. Woodward, "Regulation of Alligator Farming by the State of Florida," in *Proc. First Annual Alligator Production Conf.*, edited by P. Cardeilhac, T. Lane, and R. Larsen (Gainesville: Institute of Food and Agricultural Sciences, Univ. of Florida, 1981), 4–9; R. M. Elsey, "Louisiana's Alligator Ranching Programme: A Review and Analysis of Releases of Captive-raised Juveniles," in *Crocodilian Biology and Evolution,* edited by G. C. Grigg, F. Seebacher, and C. E. Franklin (New South Wales, Australia: Surrey Beatty and Sons, 2000).

Information describing the operations of an alligator farm was gathered in interviews of the owner and an employee of Donald Alligator Farm in Ouachita Parish, Louisiana, on April 23, 2012.

Chapter 8. A Love/Hate Relationship

A Cultured Creature

Several specifics (for example, material on books and movies) in this chapter were gleaned from the most comprehensive source on the role of alligators in human culture: Glasgow, *A Social History of the American Alligator*. I highly recommend this excellent work.

Alligator place-names can be found in F. R. Abate's *Omni Gazetteer of the United States of America* (Detroit, Mich.: Omnigraphics, 1991). Sources for alligator brand names and products are accessible by doing a web search for the specific items, for example, Alligator Records, Alligator Cable, Gatorboats, Gatorade, Univ. of Florida Gators, et cetera. The astronauts encounter with an alligator was documented by C. Moskowitz, "Astronauts Adopt Alligator as Mascot," Space.com, accessed August 4, 2012, http://www.space.com/6863-astronauts-adopt-alligator-mascot.html. The Top 10 alligator songs are from Robert the Radish, "Top 10 Alligator Songs," accessed August 4, 2012, http://music.yahoo.com/blogs/yradish/top-10-alligator-songs.html.

"Swamp People" sources include: M. Guthrie, "'Swamp People' Sets Ratings Record for History," accessed August 4, 2012, http://www.broadcasting cable.com/article/456314-_Swamp_People_Sets_Ratings_Record_for_History. php; R. Seidman (2), "History's 'Swamp People' Finishes Second Season with Record Breaking Ratings," accessed August 4, 2011, http://tvbythenumbers. zap2it.com/2011/07/22/historys-swamp-people-finishes-second-season-with-record-breaking-ratings/98703/ and "Series Premiere of History's 'Swamp People' Sets Record," accessed August 4, 2012, http://tvbythenumbers. zap2it.com/2010/08/24/series-premiere-of-historys-swamp-people-sets-record/60858/; B. Gorman, "'Swamp People' Season 2 Premiere Attracts 3.9 Million Total Viewers, Becomes History's #1 Series Telecast Of All Time For A Thursday Night In All Key Demographics," accessed August 4, 2012, http://tvbythenumbers.zap2it.com/2011/04/01/swamp-people-season-2-premiere-attracts-3-9-million-total-viewers-becomes-history%E2%80%99s-1-series-telecast-of-all-time-for-a-thursday-night-in-all-key-demographics/87972/.

The Threat—Real and Imagined

Fear of predators is discussed in the following articles: L. Y. Zanette, A. F. White, M. C. Allen, and M. Clinchy, "Perceived Predation Risk Reduces the Number of Offspring Songbirds Produce per Year," *Science* 334, no. 6061 (2011): 1398–1401; P. Prokop and J. Fancovicova, "Perceived Body Condition is Associated with Fear of a Large Carnivore Predator in Humans," *Ann. Zool. Fennici* 47 (2010): 417–25; J. W. Laundre, L. Hernandez, and W. J. Ripple, "The Landscape of Fear: Ecological Implications of Being Afraid," *Open Ecology Journal* 3 (2010): 1–7.

Mention of an alligator attack associated with the La Salle expedition is found in F. X. Martin, *The History of Louisiana: From the Earliest Period* (Gretna, La.: Pelican Publishing, 1975), 86–87. The Concordia Parish, Louisiana, incident is recorded by A. R. Kilpatrick, "Historical and Statistical Collections of Louisiana," in *De Bow's Review of the Southern and Western States* 4 (1852): 40–62.

The source of the report that lists alligator attacks between 1948 and 2005 is R. L. Langley, "Alligator Attacks on Humans in the United States," *Wilderness and Environmental Medicine* 16 (2005): 119–24. Shark attack figures are from the Florida Museum of Natural History, "International Shark Attack File," accessed August 4, 2012, http://www.flmnh.ufl.edu/fish/sharks/isaf/2011summary.html.

The material on alligator bites and fatal alligator attacks on people in Florida was found at FWC, "Alligator Bites on People in Florida," accessed August 4, 2012, http://myfwc.com/media/310203/Alligator_GatorBites.pdf. The Florida "fact sheet" is from the FWC, "Alligator Attacks Fact Sheet," accessed August 4, 2012, http://www.web.archive.org/web/20060508043857/http:/www.myfwc.com/gators/nuisance/Attack+Sheet.pdf.

Chapter 9 "What Good Is It?"

Medicine

The Leopold quote is found at the Leopold Institute, "Who Was Aldo Leopold?" (originally from *A Sand County Almanac*), accessed August 4, 2012, http://leopold.wilderness.net/aboutus/aldo.htm.

Sources for Dr. Merchant's research include: L. N. Darville , M. E. Merchant, A. Hasan, and K. K. Murray, "Proteome Analysis of the Leukocytes from the American Alligator (*Alligator mississippiensis*) Using Mass Spectrometry," *Comp. Biochem. Physiol. Part D* 5, no. 4 (2010): 308–16; C. Paddock, "Alligator Blood Could Be New Antibiotic for Superbugs," accessed

August 4, 2012, http://www.medicalnewstoday.com/articles/103400.php; Science Daily, "Alligator Blood May Put the Bite on Antibiotic-Resistant Infections," accessed August 4, 2012, http://www.sciencedaily.com/releases/2008/04/080407074556.htm.

The Swamp Canaries

Some consulted sources pertaining to contaminant issues and alligators include: R. H. Rauschenberger, J. J. Wiebe, M. S. Sepalveda, J. E. Scarborough, and T. S. Gross, "Parental Exposure to Pesticides and Poor Clutch Viability in American Alligators," *Environmental Science Technology* 41, no. 15 (2007): 5559–63; M. F. Delany, J. U. Bell, and S. F. Sundlof, "Concentrations of Contaminants in Muscle of the American Alligator in Florida," *Journal of Wildlife Diseases* 24, no. 1 (1988): 62–66; L. J. Guillette, Jr., D. A. Crain, M. P. Gunderson, S. A. Kools, M. R. Milnes, E. F. Orlando, A. A. Rooney, and A. R. Woodward, "Alligators and Endocrine Disrupting Contaminants: A Current Perspective," *American Zoologist* 40, no. 3 (2000): 438–52; A. W. Garrison, L. J. Guillette, Jr., T. E. Wiese, and J. K. Avants, "Persistent Organochlorine Pesticides and their Metabolites in Alligator Livers from Lakes Apopka and Woodruff, Florida, USA," *Inter. Journal of Environmental and Analytical Chemistry* 90, no. 2 (2010): 159–70; M. R. Milnes and L. J. Guillette Jr., "Alligator Tales: New Lessons about Environmental Contaminants from a Sentinel Species," *BioScience* 58, no. 11 (2008): 1027–36; G. P. Cobb, P. D. Houlis, and T. A. Bargar, "Polychlorinated Biphenyl Occurrence in American Alligators (*Alligator mississippiensis*) from Louisiana and South Carolina," *Environmental Pollution* 118, no. 1 (2002): 1–4; C. H. Jagoe, B. Arnold-Hill, G. M. Yanochko, P. V. Winger, and I. L. Brisbin, Jr., "Mercury in Alligators (*Alligator mississippiensis*) in the Southeastern United States," *Science of the Total Environment* 213, nos. 1–3 (1998): 255–62; G. M. Yanochko, C. H. Jagoe, and I. L. Brisbin, Jr., "Tissue Mercury Concentrations in Alligators (*Alligator mississippiensis*) from the Florida Everglades and the Savannah River Site, South Carolina," *Archives of Environmental Contamination and Toxicology* 32, no. 3 (1997): 323–28; L. A. Moore, "Distribution of Mercury in the American Alligator (*Alligator mississippiensis*), and Mercury Concentrations in the Species Across its Range," MS thesis, Univ. of Georgia, 2004, 1–60; D. C. Honeyfield, J. P. Ross, D. A. Carbonneau, S. P. Terrell, A. R. Woodward, T. R. Schoeb, H. F. Perceval, and J. P. Hinterkopf, "Contaminants, and Tissue Thiamine in Morbid and Healthy Central Florida Adult American Alligators (*Alligator mississippiensis*)," *Journal of Wildlife Diseases* 44, no. 2 (2008): 280–94; G. W. Witmer, J. D. Eisemann, T. M. Primus, J. R. O'Hare, K. R. Perry, R. M. Elsey, and P. L. Trosclair III, "Assessing Potential Risk to Alligators, *Alligator missis-*

sippiensis, from Nutria Control with Zinc Phosphide Rodenticide Baits," *Bull. Of Environmental Contamination and Toxicology* 84, no. 6 (2010): 698–702; and L. J. Guillette, Jr., D. B. Pickford, D. A. Crain, A. A. Rooney, and H. F. Percival, "Reduction in Penis Size and Plasma Testosterone Concentrations in Juvenile Alligators Living in a Contaminated Environment," *General and Comparative Endocrinology* 101 (1996): 32–42.

Index

The letter *i* following a page number denotes an illustration. The letter *t* following a page number denotes a table.

Bayou D'Arbonne, Louisiana, 2
Bear, 12, 51–52, 69, 72, 93
Beaver, 54
Big Cypress National Preserve, 56
Biloxi tribe, 65. *See also* Tunica-Biloxi tribe
Birds, 55, 62, 68, 73–74, 77; alligator anatomy compared to, 23–25, 28; as alligator food, 35, 38*t*, 40; alligator predation by, 51; as related to alligators, 8–11
Bison, 72–73
Black Bayou Lake National Wildlife Refuge, 41
Black River, Louisiana, 96
Blue-green algae, 53
Brachychampsa montana, 10
Brain, 11, 24, 26

Cabeza de Vaca, Alvar Núñez, 66
Caddo tribe, 65
Cadmium, 105
Caiman, 8, 10, 14*t*, 15, 19–20, 23
Calcium, 43
Caloosahatchee River, Florida, 73
Calusa tribe, 63, 63*i*
Cancer, 47, 106
Cannibalism, 37, 52, 104
Cat, 36, 38*t*, 61, 102
Cattail, 43
Celestial orientation, 26
Chabreck, Robert, 76
Cherokee tribe, 65
Chinchuba, 65
Chinese alligator, 14*t*, 15, 19
Choctaw tribe, 65
Chromium, 105
Chromosome, 33, 47
Civil War, American, 69–71
Climate change, 61
Cloaca, 23, 26, 42
Coastal Plain, 15–17
Cocodrie, 12, 92
Cocodrilo, 12
"Cold Night for Alligators" (song), 93

Conflicts, human/alligator, 78–79, 95–97, 97*i*, 98*t*, 99–102
Contaminant, 53, 104–5
Convention on International Trade in Endangered Species of Wild Fauna and Flora (CITES), 79–80
Copper, 105
Copulation, 42
Courtship, 41–42
Crab, 35–36, 38*t*
Crawfish, 1, 35–36, 38*t*, 46
Creek tribe, 65
Cretaceous period, 7–10
Croatan National Forest, 15
Crocodile, 8, 53, 63–64, 66, 78, 92, 96; alligator anatomy compared to, 19, 23; farming of, 89; taxonomy of, 10, 13, 14*t*, 15
Crocodilian, 12, 78; description of, 19–24, 26–27; evolution of, 7–11; taxonomy of, 14–15
Crocodilus lucius, 13
Crocodilus mississippiensis, 13
Crocodylia, 13*t*, 15
Cuvier, Georges, 13
Cypress, 3, 16–17, 56, 62. *See also* Baldcypress

Daudin, François Marie, 12–13
DDD, 105
DDE, 105
DDT, 105
Death roll, 40
Deer, 38*t*, 40, 102
Deinosuchus, 10
Dieldrin, 105
Dinosaur, 7–8, 10–11, 27
Disease, 46, 53, 65, 73, 88, 90, 104
Ditmars, Raymond, 20
DNA, 13, 33
Dog, 9, 36–37, 38*t*, 62, 100, 102
Dome pressure receptor, 24–25, 25*i*, 26
Doswellia, 9
Dysentery, 104

West Nile Virus, 53, 90

Wetlands, 3, 31, 56–57, 78, 93, 100, 105; alligator farming and, 86–87, 91; alligator movement and, 51; alligator nesting and, 1, 43; alligator range and, 15–17; humans and, 62, 74–75; prehistoric, 61–62; weather and, 53, 55

Whooping crane, 68

Wildfire, 45, 52*i*

Wiregrass, 43, 51

Woodland Culture, 62

Worm, tongue, 53

Zeitgeber, 44

Zinc, 105

Zinc phosphide, 105

KELBY OUCHLEY is a biologist who has managed National Wildlife Refuges for the U.S. Fish and Wildlife Service for 30 years. His first book was *Flora and Fauna of the Civil War: An Environmental Reference Guide*. His first historical novel, *Iron Branch: A Civil War Tale of a Woman In-Between*, was also ripe with natural history. A collection of his essays, *Bayou-Diversity: Nature and People in the Louisiana Bayou Country*, soon followed. Since 1995, Kelby has written and narrated a weekly conservation-related program for the public radio station (KEDM 90.3) that serves the Ark-La-Miss area. Kelby and his wife, Amy, live in the woods in Rocky Branch, Louisiana, in a cypress house surrounded by white oaks and black hickories. His website is http://www.bayou-diversity.com/.

THE UNIVERSITY PRESS OF FLORIDA is the scholarly publishing agency for the State University System of Florida, comprising Florida A&M University, Florida Atlantic University, Florida Gulf Coast University, Florida International University, Florida State University, New College of Florida, University of Central Florida, University of Florida, University of North Florida, University of South Florida, and University of West Florida.